The Politics of Dependency

US RELIANCE ON MEXICAN OIL AND FARM LABOR

Martha Menchaca

University of Texas Press   *Austin*

Requests for permission to reproduce material from this work should
be sent to:
    Permissions
    University of Texas Press
    P.O. Box 7819
    Austin, TX 78713–7819
    http://utpress.utexas.edu/index.php/rp-form

∞ The paper used in this book meets the minimum requirements of
ANSI/NISO Z39.48–1992 (R1997) (Permanence of Paper).

LIBRARY OF CONGRESS CATALOGING-IN-PUBLICATION DATA
Names: Menchaca, Martha, author.
Title: The politics of dependency : US reliance on Mexican oil and
farm labor / Martha Menchaca.
Description: First edition. Austin : University of Texas Press, 2016.
Includes bibliographical references and index.
Identifiers: LCCN 2015037902
ISBN 9781477309407 (cloth : alk. paper)
ISBN 9781477309995 (pbk. : alk. paper)
ISBN 9781477310007 (library e-book)
ISBN 9781477310014 (non-library e-book)
Subjects: LCSH: United States—Foreign economic relations—Mexico.
Mexico—Foreign economic relations—United States. Petroleum industry
and trade—Mexico. Mexican American agricultural laborers—United States.
Classification: LCC HF1456.5.M6 M46 2016
DDC 331.5/440896872073—dc23
LC record available at http://lccn.loc.gov/2015037902

doi:10.7560/309407

Contents

Tables

Abbreviations

CNH	Comisión Nacional de Hidrocarburos
DHS	US Department of Homeland Security
EAS	US Executive Agreement Series
EIA	US Energy Information Administration
GAO	US Government Accountability Office
IMF	International Monetary Fund

	BUFF/ED	Statements by Executive Directors at Executive Board Meetings
	CR	Country Reports, Consultation
	EBD	Executive Board Documents
	EBS	Executive Board Specials
	EBM	Executive Board Minutes
	SM	Board Document—Staff Memoranda

INEGI	Instituto Nacional de Estadísticas y Geografía
KAV	Kavass Series, US Treaties and Other International Agreements
NAFTA	North American Free Trade Agreement
NAWS	National Agricultural Workers Survey
Pemex	Petróleos Mexicanos
SEC	US Securities and Exchange Commission
SHCP	Secretaría de Hacienda y Crédito Público
SPR	US Strategic Petroleum Oil Reserve
TIAS	US Treaties and Other International Agreements

Preface

*T*HE AIM OF THIS BOOK IS TO EXPLORE THE ASYMMETRI-
cal economic codependency that binds the United States and
Mexico, two nations that share a physical border and support a common
capitalist doctrine. Theoretically, my intent is to argue that the concept of
dependency cannot solely be considered to be a characteristic that defines
a weaker nation's relationship to its more powerful partner. Dependency, I
argue, binds a wealthy nation to a poor nation as well, because the powerful
nation can become dependent on the resources of its weaker partner. Based
on my analysis of US-Mexican history, I conclude that when codependency
develops, an asymmetrical power relationship arises that disproportionately
benefits the affluent state.

My study, therefore, will explore US-Mexican codependency by examining two energy resources that are important to both nations: Mexican crude
oil and Mexican migrant farm labor. I selected these two forms of energy
because from the early twentieth century to the present, US corporations or
the US government have sought to either regulate or take ownership of them.
Both of these energy resources are sought after because they are scarce in the
United States, or insufficient. The governments and corporations of Mexico
and the United States trade many other commodities, but none of them are
as essential to the US economy as crude oil and farm labor. By examining the
commercial flow of farm labor and crude oil, I am able to illustrate how US-
Mexican economic relations evolved and became asymmetrically structured.
To ensure their flow, the US government at times enacted coercive or incentive policies ranging from imposing sanctions to extending privileged status
or providing continuous financial assistance.

In advancing my analysis, I will illustrate that US dependency on Mexican crude oil is centered upon protecting US national security interests and

reducing the cost of oil for American citizens. Historically, the United States has been the largest consumer of crude oil in the world (EIA 2013a, 2015a). This consumption level necessitates long-term economic planning to ensure that Americans have sufficient oil for their everyday needs while paying the lowest price available. Mexico, therefore, has played a major role in servicing the needs of US consumers, as it is a country with large oil deposits.

The US demand for crude oil and Mexico's ability to supply this resource generated a codependent relationship. Whereas Americans have historically needed a dependable supplier, the Mexican government in turn has relied on the United States to be its main crude oil importer. This symbiotic relationship will be examined and used to explain how Mexico's inability to maintain a stable economy has given the US government the political leverage to shape US-Mexican trade oil accords. This is a problem that Mexico has experienced periodically since the late nineteenth century, but as of 1981 it became part of normal operating procedures because the Mexican economy failed to be transformed into a first world economy. As part of US-Mexican diplomatic arrangements, most of which involved the US government issuing Mexico loans, the Mexican government was required to accept conditionality agreements affecting the value and amount of oil exported to the United States. However, I will illustrate that at times this power dynamic was temporarily disrupted, and Mexico gained independence to influence US-Mexican agreements. This generally occurred only when the US government was embroiled in an international crisis, such as during the Second World War or the 1970s Middle East conflicts.

Currently, Mexico continues to be the third-largest supplier of oil to the United States, even though oil exports to the United States have been substantially reduced since 2008 (EIA 2011, 2015b). Mexico began a conservancy program because its geological reserves began to fall and new oil deposit discoveries were insufficient to restore its daily production. The depletion of Mexican crude oil was gradual. Beginning in 1981, Mexico maintained a high crude oil production level, and accelerated it during economic crises, as oil exports were the government's main source of revenue. In 2006, Mexican geologists predicted that if Mexico did not reduce its production level, it might become an oil-importing nation within a few decades. This led the Mexican government to reduce production and explore alternatives to make Pemex (Petróleos Mexicanos) — the state-owned corporation that managed all aspects of the Mexican oil industry — a more efficient agency (see chapter 5).

In 2008 Mexico's president, Felipe Calderón, asked the Mexican Congress to consider privatizing the nation's oil industry. Calderón's proposal caused nationwide populist protests because he planned to reform Article 27 of the Mexican Constitution and reverse the spirit of the constitutional reforms of 1917 and the Expropriation Decree of 1938. In 1917, during the Mexican Revolution, Article 27 was adopted to protect Mexican hydrocarbons, including crude oil, from foreign ownership. At that time, the oil industry was nationalized, but the decree was not enforced due to US intervention. In 1938, after a long legal battle with foreign oil corporations, the Mexican Supreme Court allowed President Lázaro Cárdenas to enforce the nationalization hydrocarbon decree of 1917. He immediately terminated all foreign ownership of Mexico's oil wells and issued the expropriation decree.

Proponents of President Calderón's privatization initiative argued that selling the industry was necessary as Mexico's massive oil deposits could not be exploited because the government had insufficient capital to develop the industry. For US petroleum corporations, this was a favorable turn, as the Mexican Congress was also considering opening the bids to foreign investors if the industry was sold. Yet in 2008 the Mexican Congress, largely due to populist protests, rejected privatization, and a more moderate proposal was introduced in 2013. President Enrique Peña-Nieto then convinced Congress to support his constitutional energy reforms. On December 13, 2013, the states ratified the president's energy initiative and supported opening the oil industry to private investors. Although the oil industry was not sold, the federal government now allowed private investments, including foreign investments, in the oil and electrical industries.

Mexicans have interpreted the energy reforms of 2013 in vastly different ways. Supporters of privatization argue that Mexico is moving in the right direction. Allegedly, this is the first step to make Pemex more efficient and prepare Mexican citizens to accept the reality that the oil industry eventually will be privatized. Their vision is that under private ownership, the oil industry will become more productive and expand employment opportunities for Mexicans. Critics of the reforms also view this to be the first stage in the privatization of the industry. Their vision, however, is that Mexico must reverse policy; otherwise they foresee that US corporations will exploit the reforms and eventually pressure the Mexican government to sell the industry. This is a pattern that they allege has occurred in the past, and it will be repeated if the reforms of 2013 are not repealed. Their argument parallels an internal colony critique, which I examine in the next chapter.

It is uncertain how the Mexican people will react to the reforms in the near future, and whether they will support the status quo or demand that the revisions of Article 27 be nullified. My intent in this book, however, is not to take a political stance on this debate, but rather to chronicle the events that led to the reforms of 2013, and to offer an analysis of how US-Mexican codependency on Mexican oil served to push this agenda forward.

My analysis of the Mexican oil industry will be chronological, beginning in the late 1800s, when a US investor founded the industry, and concluding with the 2013 energy reforms. As I unfold this account, I will also turn to the history of farm labor. I will illustrate that US dependency on Mexican oil developed at the same time that American farmers became dependent on Mexican farm labor. World War II will be shown to be a critical period when this dependency was formed. The US government relied on Mexico to be a trusted ally during the war years, and as a result, a relationship that was designed to be temporary instead became a permanent structure. During World War II, Mexico exported crude oil and farm labor to the United States to help in the war effort. For Mexico, the export of these two commodities became a valuable source of income, which the Mexican government found important to continue.

My analysis of US-Mexican codependency, however, will illustrate that the Mexican government was less protective of its farm labor exchange policy than of its oversight over its crude oil trade policies. From 1942 to the present, the flow of Mexican farm labor either has not been regulated or has been insufficiently controlled by the Mexican government to ensure that workers are not mistreated or denied a living wage. I will illustrate that crude oil in Mexico has been traditionally treated as a more valuable commodity which must be protected and regulated, in comparison to the lesser efforts to protect the energy of Mexicans who try to sell their labor to the highest bidder. I associate this unbalanced state policy with a common position held by the governments of Mexico and the United States: both view Mexican farmworkers as hired hands that are disposable, are expendable, and lack the political clout to make demands upon the state. In Mexico and the United States, large-scale corporations benefit from retaining a group of people as a subclass that can be paid substandard wages. This occurs, I argue, because farmers in general seek to reduce the cost of agricultural production by controlling the wages of their labor force.

Exemplifying this lack of protective concern toward Mexican agricultural workers are the policies enacted in the aftermath of the North American Free Trade Agreement (NAFTA) of 1994. When Mexico, the United States, and Canada negotiated a free trade accord that required the eventual removal of most protective agricultural tariffs, the representatives of the three nations acknowledged this would cause massive unemployment in the Mexican agricultural industry. It was projected that small- to mid-scale farmers would be unable to compete with corporations, and that this in turn was expected to reduce agricultural employment in rural areas, because with the closure of farms, less employment would be available. NAFTA administrators projected that the benefits of this displacement outweighed the hardships it would produce. The displaced were expected to find employment in service or manufacturing jobs that were to be created by the NAFTA negotiations. NAFTA was expected to increase manufacturing employment and promote foreign investment in Mexico. Moreover, pushing out the less-efficient farmers was projected to increase food production and lower food prices. This book revisits this scenario and explores its effects upon the Mexican agricultural workers. I argue that many displaced Mexicans eventually found employment, but not in Mexico. Instead they were hired on US farms, where their labor was needed.

Furthermore, I illustrate that while Mexican agriculture was restructured after NAFTA, American farmers became increasingly accustomed to the regular flow of inexpensive undocumented labor. Hiring undocumented labor became the norm and not the exception. I do explore, however, how the Mexican undocumented labor force coexisted with a small-scale bracero program. Since 1986, on an annual basis, the US government has allowed farmers to import Mexican guest workers under the H-2A program. Nonetheless, most farmers favored employing undocumented labor because it was less expensive than hiring workers under the agricultural guest worker program. Undocumented workers constituted a less expensive labor force, because farmers did not have to pay for the import costs. Likewise, given that undocumented people worked under the shadow of the law, the US government could not guarantee that they were paid minimum wages, since such workers are often paid under the table. Under an agricultural guest worker program, labor is more expensive because the government regulates wages and requires that farmers provide housing for the workers and arrange for their round-trip transportation.

The analysis of the US farmers' evolving dependency upon Mexican labor is examined from World War II to the present. I illustrate that Mexican farm labor is essential to the US economy, not because farmers cannot recruit labor from other parts of the world but because obtaining labor from Mexico reduces costs. There are many poor people from India, Africa, Malaysia, and other parts of the world who would certainly welcome being part of a guest worker program. However, for US farmers this is not practical because it would raise labor costs and require the US government to establish a costly infrastructure to ensure that guest workers return home at the end of their contracts. On the other hand, Mexican agricultural labor is inexpensive, largely unregulated, and easily disposed of when no longer needed. I conclude with a discussion of current congressional debates over revising and expanding the guest worker program and offering amnesty to undocumented farmworker families. Since 2006, agricultural farm associations have been lobbying the US Congress to pass policies increasing the legal flow of Mexican agricultural workers and adjusting the status of undocumented workers to permanent legal residents (PLR). Farmers have repeatedly testified before Congress that Americans do not want to work in farm labor and that it is therefore necessary to import Mexican farmworkers.

In developing my analysis of US-Mexican codependency on Mexican crude oil and farm labor, chapter 1 examines dependency, modernization, and neoliberal theoretical frameworks to explore different interpretations of how dependency is manifested. My intent is to use existing literature to illustrate how dependency relations can evolve from some type of core-periphery structure and mutate into codependent partnerships. The literature that is examined offers a general overview, but it is selected to provide theoretical insight into the asymmetrical power relations that developed between the United States and Mexico. I argue that although the US government historically has exerted political pressure upon Mexico to comply with US regulations, since the 1980s the Mexican government has embraced US directives because it benefits the Mexican private sector.

Chapter 2 examines US-Mexican political relations during the nineteenth century and explores the early history of the Mexican oil industry. Its aim is to illustrate how US and British corporations came to own the oil industry in Mexico by 1910. This history provides the background to explore the role of American business in Mexico, and in general to unravel how US-Mexican economic asymmetrical relations were formed. I argue that during the Porfirian period (1870 to 1910) the Mexican government was complicit in allowing US corporations to take ownership of Mexican land, resources, and labor power. The desire of Mexico's ruling class for modernization placed the nation on a path toward economic dependency, as they believed that US capital could modernize Mexico. The political and economic relations that developed during this time parallel the core-periphery structure discussed in chapter 1. Chapter 2 also examines the reconstruction that followed the Mexican Revolution. The focus is on examining the US government's response to the treatment of American investors following the restructuring of the Mexican government. As part of this analysis I examine the oil politics that proceeded in the aftermath of the Mexican Revolution, which culminated with a US embargo of Mexican trade following President Cárdenas's oil expropriation decree of 1938. I close with the Second World War and argue that with the outbreak of the war, the US government learned that Mexico's oil industry must be allowed to flourish, as its success was a safeguard to US national security. This event generated a political process that led the US government to become dependent on the Mexican oil industry, as Mexican crude oil came to be treated as a reserve supply for US consumption.

Chapter 3 examines the transnational flow of farm labor from Mexico to the United States from 1942 to the late 1980s. It explores the development of

US dependency on Mexican labor, which began in the aftermath of World War II. It also chronicles the Mexican government's lack of involvement in ensuring that agricultural workers receive fair wages. This is largely attributed to the Mexican government's economic dependency on US financial assistance. That is, the Mexican government historically has gained considerable financial advantages by allowing the flow of agricultural workers to continue uninterrupted. The analysis of the causes of Mexican migration to the United States is further developed in the next chapter.

Chapter 4 examines Mexico's economy from 1981 to the present, focusing on periods of economic crisis when US government bailouts saved Mexico from economic ruin. It is argued that the economic crises from 1981 to 1995 triggered many events that fostered closer economic ties between the governments of Mexico and the United States. The US government, however, continued to play the dominant role and shaped US-Mexican farm labor and oil policies to benefit US interests. This period is described as a time when the two nations united for their common good. But it is also a period when American farmers became increasingly dependent on Mexican agricultural work, and when Mexicans who were engaged in agricultural occupations faced a difficult time as they were gradually displaced from this forum.

Chapter 5 focuses on the Mexican oil industry from 1979 to the present, exploring how the economic crises of the 1980s and 1990s shaped the industry's export market. This chapter also chronicles the main events that led the Mexican government to pass the energy reform of 2013. US national security interests in Mexico's oil industry are discussed as part of the evolution of Mexico's oil politics. The conclusion offers a summation of the main findings and completes the analysis of why Mexico has accepted an asymmetrical co-dependent relationship with the United States.

To recover the data for this study, I relied heavily on International Monetary Fund (IMF) primary documents (1981 to 2013), which included financial agreements between the Mexican federal government and its global lenders, as well as reports on the structural adjustment conditions placed on Mexico by the IMF, the US Treasury, and international banks. The IMF reports were used to explore the status of Mexican-US trade agreements and loan accords. I also compared this data with the accord agreements published by the US Department of State. To examine Mexico's economy, I reviewed IMF staff reports, comparing their data with World Bank statistics and the economic profiles prepared by the Mexican federal agency Instituto Nacional de Estadísticas y Geografía (INEGI). Other economic sources included reports by Bancomer, the US Accounting General Office, and Petróleos Mexicanos (Pemex; annual financial reports).

The historical and economic data of Petróleos Mexicanos is based on primary data issued by Pemex (e.g., *Statistical Yearbook*, director reports, labor studies), the Organization of the Petroleum Exporting Countries (OPEC), and the US Energy Information Administration. Concerning the history of Mexican agricultural workers, this information is based on US Department of Labor primary data and data from the US Department of Agriculture. I also compared this information with reports prepared by INEGI and the US State Department. Secondary literature was used to reconstruct the history of the migration of agricultural workers from Mexico to the United States, as well as to reconstruct the history of Mexico's oil industry. Finally, news reports and Mexican and US legislative records (e.g., congressional journals, constitutional reports, government announcements) were used to offer an overview of Mexico's current economic status, its energy reforms, and the US congressional debates over reforming US immigration policy.

Acknowledgments

*I*DEDICATE THIS BOOK TO DR. ERNESTO GALARZA, WHOSE research on farm labor and the Mexican economy inspired me to explore the politics of US-Mexican trade relations. This book would not have been possible without the financial contributions of the University of Texas at Austin. The Teresa Lozano Long Institute of Latin American Studies provided support for this research from funds granted to the Institute by the Andrew W. Mellon Foundation. This funding allowed me to visit libraries and archival depositories in Mexico City during the summers of 2011 and 2012. I also thank the Center for Mexican American Studies for the research grant I received to visit libraries in Mexico City and to review archives in Washington, DC, at the Library of Congress and the International Monetary Fund. The writing of this manuscript was also facilitated by a one-semester Faculty Research Grant from the University of Texas at Austin.

Finally, I would like to thank my brother Mauro Acosta, who encouraged me to explore the displacement of Mexican farmers after the passage of the North American Free Trade Agreement of 1994.

The Politics of Dependency

CHAPTER ONE

From Dependency to Codependency

*M*Y ANALYSIS OF US-MEXICO CODEPENDENCY BEGINS with a discussion of literature that explores theoretically how economic dependency is formed, is manifested, and evolves. These theoretical interpretations do not offer a singular vision. On the contrary, they reflect opposing interpretations of the causes that lead to dependency, and the benefits a poor nation acquires when it becomes economically dependent upon the finances and resources of a powerful nation. This chapter concludes that dependency must not be understood as an economic process that happens solely to poor nations, because relations do evolve, and in modern times they can develop to the point of entrenching rich nations in a codependent bond.[1] As it changes, the relationship may remain asymmetrical, as is the case with Mexico and the United States. It is argued, however, that when a powerful nation becomes dependent on the resources of its partner, this allows the weaker nation to negotiate favorable agreements for its citizenry.

My intent in this chapter is not to seek historiographical completeness of dependency literature, but rather to identify dominant theoretical traditions. The reviewed literature is divided into three categories: colonialism and development, the core-periphery critique, and the ascendance of neoliberal theory. I use this literature to illustrate how dependency relations evolve and may mutate into codependent partnerships. The literature examined offers a general overview of dependency theory, but was specifically selected to provide theoretical insight into US and Latin American relations.

THE EXPANSION OF EUROPE AND THE
UNITED STATES: COLONIALISM OR DEVELOPMENT?

In 1902, British economist John Hobson offered the earliest interpretation of the ecological and economic conditions that have led militarily powerful nations to expand their borders through colonization. Focusing on the British Empire, Hobson argued in *Imperialism: A Study* that in the early eighteenth century, population pressures and capitalist greed impelled the British government to embark on a colonization project across the globe. Vulnerable societies were identified, and war waged against them to force them to comply with British rule. Over the years, less brutal force was needed to maintain control of the colonies, as other social apparatuses became more effective. To maintain surveillance of the native population, the government encouraged the migration of colonial settlers, and they were given policing power over the economies and governments of the colonies. The main modus operandi to maintain control of a colony was to restructure the natives' economic structures by destroying their subsistence practices and then transforming them into consumers of colonial commodities. To prevent natives from competing with British exporters, it was necessary to prohibit the colonies from developing manufacturing industries, or competitive agricultural enterprises. Colonies were transformed into single-crop economies with most businesses owned by British investors. Free-trade policies were also imposed upon the colonies, prohibiting them from placing tariffs on imported goods. Essentially, British colonies were converted into consumer societies and destined to produce only commodities that made profits for investors.

Hobson argued that this colonial economic system was rationalized under immoral scientific logics, such as the "law of decreasing returns" (Hobson 1902: 192). That is, the underdevelopment of the colonies was rationalized as a productive process that less advanced societies must undergo to modernize. Hobson found this to be a callous excuse to defend British economic imperialism. He instead professed that capitalism could flourish in Great Britain if corporations produced fewer commodities and paid their workers higher wages. His limited production model proposed that better-paid workers would consume what the nation produced. In this way, colonizing the markets of other nations would be unnecessary. Hobson did, however, acknowledge that limited production inhibited the growth of capitalism.

In the early twentieth century, the discipline of anthropology became a leading social science to advance empirical data on the impact of European and US imperialism on the economies of colonized nations. Though anthropologists were sympathetic to the Hobson critique, this approach to understanding global connections remained marginal to the field (see Fabian

2002). Anthropologists instead advanced developmental models to explain how colonial relations were formed and evolved. It became an accepted social fact that colonization was an inevitable process that occurred when a technologically advanced nation encountered a less developed society.[2] Melville Herskovits (1938), writing within this tradition, advanced a modernization cultural continuum model. Although Herskovits acknowledged that the colonial encounter was a forced process, he concluded that eventually the conflict created by the invading nation led to the modernization of the dominated society. The colonized societies were described by Herskovits to be traditional, inhabited by peoples who believed in magic, and unable to efficiently exploit their land. To explain the relations that evolved after contact with Europeans, Herskovits argued that a modernization process followed which included the imposition of a colonial government and the reorganization of the economy. Referring to case studies advanced by other anthropologists, Herskovits argued that while the dominated groups were placed under new social regimes, a small number of them would be educated and expected to broker between the ruling government and the native people. These changes in turn allegedly stimulated the advancement of the dominated. Although Herskovits acknowledged that contact might result at first in the destruction of the native culture and the forced subordination of the dominated population, he believed this was a temporary phase. That is, during the early stages of the colony, the conquered population would be governed by the use of force. However, once the colonized adopted the culture and ways of life of the ruling civilization, preparations for self-rule began. Self-governance was projected to be achieved at least one generation after contact, a transitional phase that Herskovits named "acculturation." Herskovits hypothesized that after the colonized replicated the culture of the colonizers and acquired their technological knowledge, they entered the final cultural transition stage. This phase Herskovits labeled "assimilation." The dominated group's material culture would improve, and they would become politically independent.

Herskovits's modernization to thesis was utopian and paid little attention to the mistreatment of the colonized. The extraction of their material resources for the purpose of expanding the prosperity of the colonizing nation was ignored altogether. Herskovits, however, did advise researchers witnessing cultures in transition not to advance ethnocentric analyses concluding that cultural change was the result of the general inferiority of the dominated. By the late 1940s, most social anthropologists continued to ignore the consequences of colonialism upon third world societies and advanced various developmental models wherein they classified societies based on their level of technological advancement. The classifications introduced by anthropologists served to rationalize the actions of colonial regimes under the guise that backward nations needed benevolent modernization (see White 1949; Harris 1968).[3] By the end of the Second World War, it became a standard academic practice to culturally map societies based on their level of technological development, a state of affairs highly influenced by the research of anthropologists.

In *Encountering Development: The Making and Unmaking of the Third World*, Arturo Escobar (1995) advanced a provocative review of the use of cultural classificatory schemes in the 1940s postwar years. He argued that nations across the globe were classified as developed and underdeveloped or as first, second, and third world. Scholars used these classificatory schemes to distinguish nations based on their finances. This in turn allowed scholars to advise institutions on how to regulate global trade, design foreign aid plans, and provide appropriate investment recommendations. According to Escobar, at the end of the Second World War the victorious Allied Nations, headed by the United States, attempted to restore Europe's prewar grandeur.[4] To do so, the Allies established the United Nations and the International Bank for Reconstruction and Development (IBRD). The United Nations' mandate was to create a peaceful forum for political debate, while the IBRD was established to extend financial aid and loans to war-torn countries. As part of the bank's mandate, some financial assistance was also to be given to underdeveloped nations. To determine which nations were to receive assistance, the bank adopted classificatory developmental terms designed by scholars. Nations were differentiated according to their wealth, level of technology, and natural resources. First world nations were those that were fully industrialized, while second world nations were partially industrialized. Third world nations were not industrialized; most lacked the resources to do so and were underdeveloped. The problem with this classificatory scheme was the cultural assumptions attached to each label. For example, third world nations were described as backward, traditional, and inhabited mainly by illiterate peasants. Their lack of development was assumed to be the result of their traditional culture and not the effects of past colonial encounters. Escobar concludes that the cultural assumptions held by IBRD representatives shaped developmental theory and became the basis for determining how to best modernize third world nations.

By the 1950s, a common perception held by developmental scholars within the United Nations and universities was that the modernization of third world nations should proceed based on the plans designed by first world nations. Prominent developmental economists such as Arthur Lewis found that the best modernization approach was for first world nations to intervene in the economies of third world nations. Third world governments, if they chose to modernize, must therefore adopt the long-term plans of their investors and the advice of IBRD representatives. A common plan of action was for nations seeking assistance to reduce domestic spending and invest in their private sector. After this plan was set in motion, joint financial ventures between governments, foreign investors, and domestic capitalists would follow. Though developmental scholars projected that such a plan was bound to promote modernization, it also placed third world countries on a path toward economic dependency. Most of Latin America was classified as third world, with countries such as Brazil, Mexico, and Argentina considered underdeveloped countries with higher potential. The latter nations were found to have valuable natural resources and higher national incomes, yet they continued to be labeled third world because they languished between being traditional and modern.

In 1954, Lewis popularized the dual economy modernization model, which was adopted by the IBRD and many developmental scholars. Lewis proposed that third world governments, especially those with potential, could rapidly improve their economies by accelerating their industrialization. To achieve this, governments must institute projects to convince agricultural workers to abandon farming and relocate to cities, where industrial jobs were available. A key variable in projecting the success of an industrialization project was to convince workers to accept near-subsistence wages. Paying low wages was essential, as this would allow employers to accumulate adequate savings and reinvest the surplus capital to expand industrial employment. To persuade agricultural workers to abandon farming, Lewis proposed that employers must offer wages 20 percent above what they currently paid.

During the 1950s, Argentinean economist Raúl Prebisch (1950) offered a different modernization approach. He also contested the perspective that underdevelopment was the result of the backwardness of a people. Using the case of Latin America, Prebisch instead argued that the underdevelopment of most countries was caused by a history of asymmetrical trade relations. His perspective on development was influenced by Keynesian economics, which proposed that a stable and prosperous economy is based on a fair exchange of wages for labor. To maintain fair exchanges, governments must regulate labor and corporations to ensure that economic activity remains productive. Thus, Prebisch proposed that Latin American industrialization must not aim solely to produce commodities that substitute for imports; it must also aim to "increase the measurable well-being of the masses" (Prebisch 1950: 6).

In 1950, when Prebisch was invited by the United Nations to write an economic report on the status of Latin America, he delineated how a history of asymmetrical trade practices with Europe and, particularly, the United States had stagnated Latin America's industrialization. The report was titled

The Economic Development of Latin America and Its Principal Problems. It became the first study to identify the procedure used by first world nations to manipulate Latin American trade. Prebisch focused on trade and, strategically, did not alienate sympathetic supporters by dwelling on Latin America's European colonial past. A few years later his analysis influenced the development of popular theoretical schools of thought, including the core-periphery perspective and import substitution models.

In this report, Prebisch argued that US corporations had the power to control Latin American trade because they were the region's principal importers. The problem for Latin American countries was that US firms were only interested in purchasing raw materials, and not manufactured goods. Exacerbating matters, US corporations set the purchase value of the raw materials very low. According to Prebisch, Latin American governments had to accept these trade policies; otherwise, US corporations would do business elsewhere. This created a serious trade deficit problem in Latin America since the price of US manufactured goods constantly rose, while the price of raw materials either remained the same or fell. Latin American countries therefore accrued substantial foreign credit debt when financing imports. Although Prebisch identified asymmetrical trade relations as a hindrance to development, he believed that this scenario could be reversed. He optimistically projected that US corporations would open their markets to Latin American manufactured commodities after they completed the postwar reconstruction of the European markets. At this time, US corporations concentrated most of their trade with Europe in order to restore their allies' markets.

To resolve Latin America's trade deficit, Prebisch also recommended that governments adopt and finance import substitution. Latin American governments working with domestic elites were to develop an economic system in which they consumed what they produced. This entailed transforming their economies by expanding the manufacturing sector, converting subsistence agriculture to industrial agriculture (domestic/export), and modernizing transportation infrastructures to ease the movement of goods. To do this, Prebisch proposed, Latin American governments must begin by purchasing from the United States the machinery and technology to expand their own manufacturing plants. To ensure that Latin Americans consumed what their factories produced, it was essential to reduce imports. For this plan to succeed, it was also critical for Latin American governments to convince US firms to purchase Latin American commodities. Prebisch acknowledged that this plan might be difficult to institute because although US corporations most likely would be willing to sell the machinery, they would not necessarily purchase Latin American goods. To overcome this problem, Prebisch recommended developing a long-term Latin American multilateral trade and industrialization agreement. The industrialization of each Latin American country would be coordinated to avoid excessive competition and prevent the production of surplus goods that could not be consumed within Latin America.

Other Latin American economists echoed Prebisch's recommendations a few years later. In 1965 the United Nations commissioned several Latin American economists to write a report on the status of Latin American trade. The resulting "Economic Commission for Latin America" came to be known by its acronym — CEPAL (Escobar 1995; Prebisch 1971). The commission offered an extensive critique of US–Latin American trade. Prebisch was one of the commissioners. In the report the commissioners adopted several key concepts developed in Prebisch's 1950 United Nations report. Using Prebisch's concept of "center and periphery," the commissioners affirmed that US–Latin American asymmetrical trade exacerbated the continent's underdevelopment. The United States was described as the central and controlling partner. Latin American countries were described as peripheries, which were pressured to accept US directives. CEPAL described the same trade problems that Prebisch had previously noted: US corporations were only willing to import raw materials and refused to purchase any commodity that competed with US manufactured goods. When US industrialists invested in Latin American manufacturing, they generally did so in areas that did not interfere with the sale of US exports. In essence, the commission concluded that US corporations preferred Latin American nations to remain dependent on US imports.

In 1971 the Inter-American Development Bank commissioned Prebisch to prepare a second report on the status of Latin American economic development, this time covering the years 1958 to 1965. The Organization of American States, which included the United States and twenty-one Latin American nations, established the Inter-American Development Bank in 1959. The bank's aim was to promote the economic development of Latin America and, later, the Caribbean. The United States owned the majority of the bank's stock. Prebisch once again identified the two most critical issues affecting Latin America's future development: the need to institute better import substitution, and the challenges to be overcome in gaining US support. He reemphasized the point that to reduce foreign debt acquired by overimporting US manufactured goods, Latin American governments would need to finance import substitution. Latin America's predicament, however, was that its national governments needed to incur more foreign debt to finance import substitution because most Latin American nations did not have the capital to establish manufacturing plants. It was therefore critical for Latin American countries to finance import substitution without losing control of their long-term industrial plans. Identifying Mexican industrialization initiatives as possible models, Prebisch stated that from 1958 to 1965 Mexico had implemented two policies to stimulate industrialization: providing grants to improve agricultural production, and retaining a dual economic agrarian system. In Mexico, industrialization had been financed without disturbing the productivity of the agrarian sector. Mexico's long-term plan was to gradually modernize agriculture, at the same pace that occupations in manufacturing plants were created. That is, if Mexico was to become a leading industrial exporter, it would need to convert a part of the agrarian population into industrial workers and consumers of the commodities they manufactured. This plan had to be implemented gradually, or the productivity of the traditional sector would fall, erasing the gains made by the industrial sector. Import substitution projects therefore were being implemented in both urban and agrarian areas.

To modernize the agricultural sector, the government implemented projects assisting mid-scale and large-scale farmers to improve productivity. Farmers were issued credit, roads were built near their farms, irrigation improvements were subsidized, and technical expertise was provided. These investments were necessary as only these types of farmers had the capacity to produce large amounts of food to fulfill the anticipated consumption growth of the urban population. However, ensuring that the traditional sector remained productive was also important, as subsistence farmers needed to provide for their families and communities. In the traditional sector, the government financed crop assistance projects for *ejido* farmers (communal farmers) and offered technical advice.

Prebisch (1971: 29) estimated that 41 percent of Mexico's working-age population was engaged in agricultural occupations when the government implemented a plan in 1965 to gradually transition part of the youth workforce into manufacturing employment. Workers from the countryside were being recruited at a rate that could be absorbed within manufacturing, although manufacturing jobs were expanding at twice the rate of the number of workers available. The government was aware that a rapid occupational transition was best for the industrial sector because an oversupply of labor would be created and cause wages to fall. For the nation as a whole, however, rapid industrialization was counterproductive because all gains made in the industrial sector would be lost in the traditional sector if productive youths were converted into unemployed laborers.

By 1968, Prebisch had concluded that Mexico's long-term import substitution plan was working. Manufacturing profits were up, and agrarian youth were being gradually transformed into wage workers. The productivity of the agrarian sector was also on the rise. Mexico's GDP had risen to 6.3 percent, one of the highest in Latin America (Prebisch 1971: 49). More significantly, Mexicans were consuming what was produced in the country, and imports had declined. Prebisch also found that Mexico had reduced its foreign debt, as fewer loans were needed to pay for imports. In 1970 Mexico's total foreign debt was $9 billion in US dollars (Martínez Fernández 1996: 18–19).

If other Latin American countries duplicated Mexico's import substitution plans, Prebisch projected that they could move toward economic independence. This, however, was highly unlikely, since Mexico had the capital to finance its import substitution, while most other Latin American countries did not. Without foreign aid or loans, Prebisch concluded, most Latin American countries could not industrialize. It was therefore necessary for Latin American governments to apply for assistance without losing control of their factories to foreign investors; otherwise, the goals of financing import substitution would fail and would primarily benefit outside interests.

THE CORE-PERIPHERY MODEL:
A RADICAL CRITIQUE OF DEPENDENCY

Although Raúl Prebisch and the CEPAL commissioners were the first to introduce and develop the economic concept of "center and periphery," their position was radicalized by other scholars. The center and periphery model shifted the analysis of the outcome of development theory from a position concluding that dependency resulted in the modernization of poor nations to one arguing that the underdevelopment of poor nations would worsen over time (see Frank 1967). In 1974, Immanuel Wallerstein's core-periphery adaptation of Presbich's theory, known as the "world systems model," reached international prominence. Wallerstein argued that when one region or state is economically prosperous, it will try to dominate nearby regions in order to expand its economic resources and political territorial control. The area of high growth becomes known as the core, and the neighboring areas the periphery. Cores and peripheries can be towns, cities, states, or nations. The initial expansion of the prosperous nation or region is generally the result of innovations in technology that allow it to accumulate surpluses beyond basic subsistence needs. Those who govern the core region distribute part of their surpluses among loyal members who live in the core, and this engenders trust, compliance, and loyalty among the governed. When a governing body establishes bonds of trust with its membership, the leaders gain the ability to mobilize their supporters and expand their political territory through colonial processes, including launching wars of conquest. Wallerstein also argued, however, that the expansion of a core nation can be the result of circumstances that may have weakened the periphery regions, such as continuous periods of famine, ecological disasters, or a society's inability to protect itself from an enemy invasion.

Like Hobson had argued in 1902, Wallerstein proposed that the colonization of vulnerable regions leads to the labor exploitation of the colonized and their failure to control their own resources. Furthermore, because an economic structure is imposed on the colonized, over the years the dominated region becomes dependent on the laws, intellectual expertise, and finances of the invading society. The economy of the periphery is strategically developed to serve the needs of the core, which generally entails extracting raw materials from the periphery (e.g., oil, gold, crops, minerals, hides) to enrich the core. The periphery also serves as a market for manufactured goods and other surpluses that cannot be consumed in the core due to overproduction. Wallerstein found that one result of the core-periphery relationship is that two-way migration becomes a structural component of the economy and political structure for both the core and the periphery. To retain political control of the periphery, the colonial government sends representatives to govern the colony and likewise permits colonized people to migrate to the core. The outmigration of the colonized is generally stimulated by lack of employment in the periphery, or when cheap labor is needed in the core nation. Migration therefore is a key element of the relations that develop.

Studies on Latin America also contributed to the development of the core-periphery model. In 1967, Andre Gunder Frank explored the role Latin American elites played in protecting the economic structures established by colonizing nations. Frank argued that core-periphery relations generally begin with a war of conquest or some type of violent intrusion by a colonial power. Relations of power evolve over time and mutate into the economic dependency of the periphery. Eventually, when decolonization takes place, elites from the periphery become complicit in supporting the continuous economic intervention of the colonial state. Frank proposed that in Latin America at the end of Spanish colonial rule in the nineteenth century, elites were instrumental in developing new core-periphery relations with the United States and resuming the exploitative economic networks that tied Latin America to Spain and Europe. By 1950, Latin America's most powerful political elites had forged business partnerships with foreign corporations, and their projects did not benefit the welfare of the common person or lead to the development of domestic industries. In essence, Frank argued that in modern times it is the national elites who convince their governments to accept business ventures and foreign trade agreements that benefit only the ruling class.

Frank also argued that in the aftermath of the Second World War, foreign aid given to Latin America made countries more dependent on the United States and stagnated their economic development. The United States, the World Bank, and the IMF offered assistance to Latin American countries, but often in the form of high-interest loans. Loan servicing and debt amortization in turn strained the national economies, since a large percentage of the federal budget was reserved to meet quarterly payments. Many countries exacerbated matters by entering a vicious debt-bondage relationship with their creditors. They continuously borrowed new money to pay for previous loans. By 1961, most Latin American governments were borrowing at a 16 percent interest rate and had little capital to invest in their nation's industrialization.

In 1977, economist Fernando Henrique Cardoso, a former member of CEPAL, concurred with Frank that most Latin American countries were peripheries of the United States.[5] In an influential article titled "El consumo de la teoria de la dependencia en los Estados Unidos," Cardoso explored Latin American dependence on international markets. He argued that on the one hand, countries such as Brazil, Mexico, and Chile had exerted greater control over their economic development owing to the revenues they collected from the exports of valuable raw materials (e.g., oil, minerals, natural gas). Most other countries, on the other hand, mainly relied on US investments; consequently, their industrial development was limited to the business needs of American investors. In general, only low-paying labor-intensive manufacturing industries (e.g., textile, toys, furniture) were established in such countries. According to Cardoso, foreign firms established factories in Latin America to reduce labor costs and gain a competitive edge in the global market. To receive and renew contracts, Latin American governments were pressured not to raise their nation's minimum wage.[6]

THE DESCENT OF THE CORE-PERIPHERY MODEL
AND THE ASCENT OF NEOLIBERAL THEORY

In *Towards a Socio-Liberal Theory of World Development*, Arno Taush (1993) posited that in the early 1990s the core-periphery model lost its prominence, as many scholars concurred that decolonized countries damaged their own economies by failing to institute democratic systems giving the masses political say in the development of their nation. Citing Latin America as a case in point, Taush claimed that many countries failed to industrialize because the old ruling elite, who often forged alliances with the military, preferred to retain the old system of dependency. For a developing or underdeveloped nation to modernize, Taush proposed, it must adopt the socio-liberal political system of first world nations. This entails giving social mobility to all social classes by instituting a democratic system. Governments must legislate fair wage laws; provide widespread public education; grant farming subsidies; implement a fair taxation system that taxes income, wealth, and inheritance; and institute a legal system that represents the political rights of all members and not only the elite. Nations that adopt social liberalism will be better equipped to stabilize their political systems and develop prosperous economies, as citizens become more productive when they are literate, are healthy, and have diverse skills. Moreover, a nation becomes equipped to protect its resources from predatory nations only when its citizens are educated and are given the power to elect representatives of their choice. In essence, Taush argued that the means to end dependency is social liberalism, because nations shape their economies according to the needs of the people rather than those of the former colonizing nation.

Taush also proposed that decolonized nations with valuable natural resources have the ability to reshape their dependent relationships by instituting import substitution. In particular, nations with large oil deposits have the resources to finance development projects; they can reinvest the profits gained from drilling oil. In the early execution of import substitution policies, raising foreign tariffs in sectors where industries are being developed is critical to protect domestic companies while they acquire the capacity to compete with foreign trade. Taush cautioned, however, that heavily indebted nations could not implement import substitution projects because their foreign creditors will oppose any policy that raises tariffs and reduces exports.

Arturo Escobar (1995, 2012) also concluded, as Taush had, that by the mid-1990s radical core-periphery models had lost their eminence within academic circles because the politics of global trade were more complex. This shift, however, also altered opposing theoretical frameworks proposing that third world dependency was caused by the backwardness of the debtor nations. Related arguments claiming that colonialism had benefited many countries also lost credibility since throughout the world the economies of third world countries had failed to improve when under foreign control. Escobar concurred with Taush that the causes of underdevelopment were multiple and must be examined from both domestic and international perspectives. But he did not endorse the view that core-periphery models were no longer theoretically applicable. In the case of Latin America, Escobar argued that trade continued to follow the core-periphery structure. Escobar nonetheless did concede that alternate models might better elucidate the experiences of Latin America's oil-exporting nations, since such countries have the resources to negotiate trade agreements with US corporations. Neo-extractivist models could better explain the experiences of oil-exporting nations (Escobar 2012). For example, in Venezuela and Ecuador, the presidents depend on the voice of the people to adopt laws prohibiting their governments from negotiating accords unfavorable to the nation. In these countries, first world nations continue to solely be interested in extracting raw materials such as oil, yet a system has been developed to ensure that a higher percentage of the value of the extracted product remains at home. Escobar concludes that in this modern era, export-import trade politics continue to be the most important issue shaping Latin America's economic future. The dilemma for Latin American governments is how to protect their markets in an era when neoliberal beliefs shape global trade policies.

NEOLIBERALISM: A DOCTRINE OF THIRD WORLD ECONOMIC DEPENDENCY OR INDEPENDENCE

Neoliberalism is an economic philosophy adopted in the 1980s by many first world nations. In the early 1990s, scholars began to document how neoliberalism began to spread across the globe. One academic tradition proposes that neoliberalism is a doctrine about how domestic and international markets function, while the other tradition treats it as the new rationale for neocolonialism, merely cloaked in new garb. The latter tradition argues that neoliberalism is a doctrine exported by first world nations and imposed upon third world economies. Its purpose is to protect the economy of first world nations and keep the international markets of developing and underdeveloped nations dependent on their finances. In essence, given that we are living in a neoliberal historical phase, scholars are in the process of debating its benefits and the outcome of neoliberal experiments.

Critics of neoliberalism argue that governments throughout the world began adopting neoliberal principles in the early 1980s largely due to pressures imposed by first world nations (Flores-Quiroga 1998). This was a period when the economy of industrialized countries slowed down, world trade was sluggish, inflation was high, and oil price warfare between OPEC and first world countries forced developing countries to form alliances with their benefactors. This global situation promoted the rejection of social liberalism as the popular doctrine of governance and the adoption of neoliberalism as an alternate approach to improve world economies. Decreasing regulation of corporations became a key element of this new doctrine, as government intervention in the marketplace was seen to impede the productivity of the private sector. To promote the prosperity of the private sector, the state had to transform its relationship with its citizenry. Less government intervention in private life came to be seen as a necessity, with individuals expected to take personal responsibility for their basic needs and not expect the state to intervene between them and the corporations. The decreasing role of the government in the lives of the people was fundamental in creating an efficient and productive citizen. When people took responsibility for their lives, the finances of the state were reserved to stimulate the economy rather than being spent in programs that did not generate wealth. Only education and health services were essential entitlements, since the private sector required healthy workers and people with varied skills to advance a free market system.

Cultural geographer David Harvey (2006) posits in "Neoliberalism and the Restoration of Class Power" that neoliberal policies throughout the world have taken different forms, with their shape depending on the power that citizens hold to make governments and corporations accountable to them. General patterns found in states that have adopted neoliberalism as a doctrine of governance include: decreased public spending, deterioration of labor unions, privatization of state agencies, lowering of protectionist tariffs in non–first world countries, tax reforms favoring corporations, and deregulation of laws pertaining to the environment and the private sector. Although I concur with Harvey's assessment, I must add that when neoliberalism is exported to developing nations, it unfolds upon preexisting political and economic structures. Neoliberalism could not develop in the same manner in Mexico as in the United States, because Mexico's standard of living historically has been considerably lower. More significantly, prior to the neoliberal turn, Mexico's social liberal policies guaranteed Mexicans few economic comforts in comparison to the policies of industrialized nations such as the United States. Social liberalism in industrialized nations, prior to the neoliberal turn, has been characterized as a form of governance that ensured its citizens a stable monetary system in which the majority could enjoy a high standard of living. Widespread infrastructure and public education were also standard practices that were not limited to regional areas, such as in the case of developing and underdeveloped nations. Fair wage laws and union representation were also accepted practices, allowing workers to negotiate better wages. Thus, when neoliberalism was introduced in Mexico, it was imposed upon an economic and political system that offered its citizens low wages and a low standard of living; consequently, with the shift in philosophy, the government's responsibility toward its people further deteriorated. In essence, neoliberalism did not cause Mexico's economic problems; it mainly restructured the government's responsibility toward its citizens. Neoliberalism, however, did lead the Mexican government to protect and privilege the private sector, even when the policies enacted went against the financial interests of the ordinary Mexican citizen. As will be discussed in chapter 4, deregulation and privatization of state agencies from the 1980s to the mid-1990s mainly served private sector interests. The outcome of these fiscal management changes led to economic turmoil and forced the Mexican government to reduce public spending in order to finance the foreign debts corporations incurred in association with the restructured economy.

Like Harvey, Arturo Escobar offers a harsh critique of neoliberalism, proposing that it is a new form of imperialism imposed upon the economies of non–first world nations (Escobar 2012, 2010). Neoliberalism in Latin America has led nations to remove most protective tariffs and open their markets further to US exports. In most cases the removal or lowering of tariffs has been unidirectional and part of trade concessions Latin American countries are pressured to accept if they are to continue conducting business with US corporations. Ultimately, this relationship is one in which Latin American countries are given a "take it or leave it" choice. Focusing on Mexico, Mexican economists Rolando Cordera and Carlos Tello (2010) illustrate how the scenario theorized by Escobar unfolds. As director of the Banco de México under President José López Portillo in the early 1980s, Tello personally observed the introduction and application of neoliberal policies in Mexico. Cordera and Tello argue that first world nations control international trade by imposing conditions on their third world partners. Before third world nations are given economic aid or are selected to be trading partners, they must accept structural adjustment agreements, which generally contain clauses removing trade barriers to foreign exports. The outcome of this process is the creation of economies that become dependent upon first world trade and investments.

Scholars who support neoliberalism as an efficient form of governance argue that open markets benefit the most competent capitalists irrespective of their nationality, lead to lowering of prices, and make commodities—specifically food—more abundant and accessible to the masses (Edelman and Haugerud 2004). Under neoliberalism, deregulation and support of the private sector are essential if the economies of nations are to move from stagnation to prosperity. Edelman and Haugerud argue that neoliberalism should not be treated theoretically as an "evil twin," as this market approach did not create or perpetuate the poverty many middle- and low-income nations experience. They also propose that scholars should recognize the positive aspects neoliberalism has introduced across the world. With the opening of markets and removal of trade barriers, first world nations have been actively involved in improving the infrastructure of developing nations by funding non-governmental organization (NGO) projects. NGOs are nonprofit organizations that are not owned or managed by governments. The organizations are dedicated to promoting human rights, protecting the environment, and engaging in other forms of charity aiming to help the poor or to improve the lives of disenfranchised peoples, specifically undocumented immigrants.

NGOs funded by foreign nations flourished in third world nations after free market policies were instituted (Appadurai 2000).

Aihwa Ong (2007) concurs that when the markets are liberalized, foreign corporations finance development projects in third world nations. Many of the beneficiaries are women and immigrants who need financial assistance to establish businesses or educational initiatives. In communities where NGOs are established, improvements in the local population's standard of living usually take place. The problem that Ong observes, however, is that corporations have benefited much more than most individuals from neoliberalism. The assets corporations accumulate via exports or foreign investments are much more lucrative than the finances poor communities receive by way of charitable projects.

Using the case of Mexico, Cordera and Tello (2010) offer a similar observation of the disproportionate benefits that neoliberalism bestows on different sectors of a society. Large-scale private corporations generally benefit much more than the public does. For example, under the neoliberal presumption that large corporations are better equipped than the government to generate employment, it has become a common practice in Mexico for the federal government to bail out large-scale corporations when they are in financial trouble. During the 1980s and 1990s, the government repeatedly assumed defaulted private sector loans and converted them to public debt. The International Monetary Fund, one of the main global institutions promoting the philosophy of neoliberalism, acknowledged in several reports that in the mid-1990s this was a serious problem in Mexico, stagnating the nation's economic growth (Baldacci, de Mello, and Inchauste 2002; Corbacho and Schwartz 2002).

FIRST WORLD CODEPENDENCY

The concept of dependency as reviewed in the aforementioned literature has languished between discussions framing economic dependency as either the natural outcome of market competition between traditional and modern economies or resulting from the historical legacy of colonialism. Theorists who claim that economic dependency is mainly a product of colonialism or neocolonialism have also argued that elites from third world nations have contributed to this dependency by mismanaging their nation's finances (see Cordera and Tello 2010). In such cases, the government, which is controlled by elites or by an authoritarian regime, follows macrostructural economic long-term plans that benefit only the governing elite and the corporations they own.

My interest in exploring dependency theories is not to advance a judgment of the debate, but rather to examine how dependency relations affect first world nations. My intent also is not to dismiss the valuable theoretical insights scholars have advanced concerning how decolonized nations respond to the pressures imposed upon them by powerful nations. By examining US-Mexico trade relations concerning oil and farm labor, my aim in this book is to argue that a powerful nation does become dependent on the resources of its weaker trading partner, because it becomes accustomed to the financial benefits it has historically gained from an asymmetrical relationship. Asymmetry, however, can entrap the dominant partner, because it establishes a dependency on the resources of the less powerful nation. Once that dependency is formed, the weaker nation has leverage in limiting or manipulating the demands of its powerful trading partner.

In the case of the United States, I concur with Arjun Appadurai (2000) that since the early 1970s, wealthy nations have not been able to employ the same techniques to control the economies of developing or underdeveloped nations. Appadurai attributes this to two transformations prompted by the effects of globalization: the rise of human rights social movements, and the transnational flow of the media. Appadurai argues that capitalism in theory continues to be characterized by "strategies of predatory mobility (across both time and space) that have vastly compromised the capacities of actors in single locations even to understand, much less to anticipate or resist, these strategies" (ibid., 16). But although one cannot deny that in many places foreign corporations continue to exploit people in the search for profit, Appadurai proposes that at the same time social movements to advocate for the poor have risen across the world, and often are supported by capitalists who favor improving the lives of the poor. He attributes the rise of compassionate capitalism and human rights social movements to a philosophical disjuncture with dominant views over the notion that poverty is self-perpetuated, is natural, and cannot be blamed on capitalist greed. Global dissent against condoning the exploitation of the poor has led to the international formation of civil networks that advocate equity, justice, and fair governance. In essence, Appadurai suggests that an international civil society exists that maintains a close watch on corporations and discloses their predatory actions to the public.

To Appadurai (2000, 2006), technological advances in establishing global media networks were pivotal in transforming the economic relations between first and third world nations, because in modern times, news travels quickly across the globe. By way of the media, citizens immediately learn about the actions of governments and corporations across the globe, giving people the opportunity to pass judgment on international affairs and, if necessary, demand remediation of unjust practices. In a sense, Appadurai's analysis reverses Michel Foucault's concept of "surveillance," as technology gave people the power to maintain a panoptic surveillance of the state. Due to technological advances in communications media, the actions of first world corporations and governments have become transparent, and powerful agents must consider the consequences of their actions. This has given third world nations the power to negotiate and possibly to limit the demands placed upon them by foreign governments and corporations.

Although I agree with Appadurai's compassionate capitalism analysis, I reflect upon the ideas of Raúl Prebisch when considering the case of Mexico. Import substitution in Mexico may be part of the blueprint of past and long-dead visions of forging a prosperous form of capitalism, but its philosophical principle remains alive. As Prebisch argued, nations such as Mexico that possess valuable resources acquire the capability to develop administrative structures to obtain the highest exchange value. As will be discussed in the following chapters, Mexico has many valuable resources, but only oil and farm labor have become essential to US consumers. US dependency on these resources has given Mexico the economic leverage to negotiate the exchange process and in turn influence how the asymmetrical relationship that has bonded both nations evolves. US dependency on Mexican oil and farm labor thus no longer makes dependency unidirectional.

CHAPTER TWO

The Politics of Oil and National Security

THE BEGINNING

*T*HE GOVERNMENTS OF THE UNITED STATES AND MEXICO share an economic partnership that has been shaped by a common border and the belief that capitalism is the greatest wealth-producing system fueling a nation's prosperity. This capitalist relationship began in 1848, when the US government sought new land for national expansion and in the process took one-third of Mexico's territory (Takaki 2000). Scholars have depicted the expansion of the United States into Mexico's northern frontier, and the early relations that developed between both nations, in terms associated with a core-periphery structure (Blauner 1994; Cordera and Tello 2010; Prebisch 1971). Under this structure, when one region or state expands due to its economic or military prosperity, the region of high growth becomes the core and the politically engulfed the periphery. In the case of US expansion into Mexican territory, this process was further facilitated by Mexico's weak political position. In 1821 Mexicans had successfully won their war of independence against Spain, but years of armed struggle then left the economy devastated and vulnerable to a military invasion.

In subsequent time periods, social class conflicts in Mexico contributed to the formation of economic alliances between Mexican elites and foreign investors, which had an overwhelmingly adverse effect on the welfare of the common Mexican citizen. In particular, social class stratification during the Porfirian period of 1872 to 1910 came to resemble a core-periphery political alliance. Mexico's ruling class opened the door to US corporations and allowed Americans to treat Mexico's most valuable resources as property belonging to the United States. US corporations took possession of Mexican oil, minerals, and land with the aid of Mexico's elites.

In this chapter, I will use the case of Mexico's oil industry to illustrate how Mexican elites during the Porfirian period became economically dependent on US investments. The desire of Mexico's ruling class for modernization placed the nation on a path toward economic dependency based on the belief that only foreign capital, specifically US investments, could transform Mexico. Mexico's economic dependency on US capital, however, was two-sided. US corporations certainly benefited from the raw materials they extracted, but at the same time they began to forge an economic relationship in which Americans became dependent on inexpensive Mexican crude oil and, by the end of the Second World War, cheap farm labor as well.

This chapter will also begin to unfold the argument that during critical political moments, US national security interests have prompted dependency on Mexican resources. I now turn to an analysis of the first stages of US-Mexican asymmetrical economic codependency. I begin with the Porfirian period, when capitalist investors began drilling oil wells in Mexico. The chapter concludes with the Second World War era, when the US government realized that in the best interests of national security, US oil corporations should share part of their profits with Mexico and not impede the development of its oil industry.

UNITED STATES TERRITORIAL EXPANSION: AMERICANS WANT LAND, NOT A COLONY WITH PEOPLE

The governments of Mexico and the United States share a long history of international relations dating back to the formation of the US-Mexican border in 1848, when the US government through war acquired Mexico's northern territories (Menchaca 2011). The US government did not plan for Mexico to become a colony; it merely wanted Mexico's land. Mexico's northern frontier had a well-developed road system connecting Mexico City to what today are the southwestern states of California, Texas, New Mexico, and Arizona. The roads in turn joined towns, villas, garrisons, missions, and ranches scattered throughout the Southwest.

The Mexican-American War of 1846 to 1848 brought closure to a long-standing land conflict that had begun in 1845, when the US government acquired most of Texas through annexation. Texas had been part of Mexico since the seventeenth century, but after US and European immigrants defeated Mexico in the Texas Revolution, Mexico ceded central and northern Texas, and the Republic of Texas was born in 1836. Nine years later, after prolonged negotiations with the US government, Texans chose to become citizens of the United States, allowing Texas to enter the Union as its twenty-eighth state.

Mexico lost more territory to the United States in 1848 after its defeat in the Mexican-American War. With the signing of the Treaty of Guadalupe Hidalgo on February 2, 1848, Mexico ceded California, New Mexico, northern Arizona, the El Paso valley (southwest Texas), and South Texas.[1] Mexico also lost parts of its northern frontier that include the modern US states of Nevada, Utah, parts of Colorado, and small sections of Oklahoma, Kansas, and Wyoming. While the treaty was being deliberated, the US Congress had the opportunity to take all of Mexico (see *Derecho Internacional Mexicano* 1877). Instead, it chose only to take the least populated territories, to avoid serious constitutional debates over race and citizenship.

Mexico was a large nation that did not fit the American binary system of law, which granted US citizenship only to whites. If Mexico were annexed, the size of the nonwhite population within the new US borders would more than triple and potentially produce serious constitutional problems because many nonwhites would have to be given US citizenship. Giving citizenship to Mexicans of unmixed Spanish descent, and denying citizenship to Mexican Indians and those with black ancestry, would not disrupt tradition or be in conflict within US law. Yet denying citizenship to racially mixed light-complexioned Mexicans who practiced Spanish culture and were not tribal peoples was a problem (Romero 1889).

Entering a national constitutional debate over which Mexicans could be treated as "white" and considered eligible for US citizenship could be avoided if only a small percentage of this population was incorporated. Once the southwestern states were annexed, Congress gave the legislatures of the annexed territories the power to determine which Mexicans were to be given citizenship. However, under federal law Congress barred blacks and tribal Indians from obtaining US citizenship. The 1854 US census bureau estimated that before the war, approximately sixty thousand Mexicans had resided in the occupied territories of California, New Mexico, and Arizona (US Census 1854: 39). No demographic estimate was given for Texas, and it is uncertain how many Mexicans there were denied US citizenship.

Southern Arizona became part of the United States in 1854 after president Antonio López de Santa Anna sold this territory under the Gadsden Purchase. At this time, Mexico was undergoing political restructuring and about to embark upon *La Guerra de la Reforma* (Reform War) after Benito Juárez, the leader of the Liberal Party, challenged the presidency of Santa Anna. President Santa Anna, in need of funds to finance a military attack against the political coalition formed by the Liberal Party, agreed to sell southern Arizona to the United States (Kluger 2007).

The seizure and purchase of Mexican property by the US government fostered Mexican resentment and hostile relations between both nations. However, an era of warm diplomatic relations commenced in 1868 when Juárez, then Mexico's president, and US president Abraham Lincoln set aside the ill feelings caused by the Mexican-American War and began to negotiate binational agreements to bring order to the movement of people and commodities crossing their border (see Menchaca 2011). An era of good neighbor relations had commenced after both governments gained victories in the mid-1860s—the US government in the US Civil War and Mexico over the French alliance that had temporarily invaded Mexico and placed Archduke Ferdinand Maximilian as emperor of Mexico.

Following the defeat of the Confederacy and the French, President Lincoln and President Juárez enacted economic treaties to stimulate international trade. Juárez needed funds to finance a development plan to improve Mexico's infrastructure, which had been partly destroyed during the French invasion and the Mexican-American War. For their part, US businessmen were interested in extracting gold and silver from mines that Mexico did not have funds to exploit. Mexico was willing to allow US corporations to lease land and extract minerals that were to be exported to the United States for a minimal federal fee. Corporations were to pay the federal government an annual fee of three hundred thousand pesos per mine (Scholes 1957: 170). The federal government was also to receive 5 percent of the mines' profits and levy taxes of 1 percent of the worth of the gold exported and 8 percent of the silver. Except for the Federal District (Mexico City) and the state of Baja California, a state government could not levy a state tax. Besides the tax revenue, President Juárez believed that US corporations would create employment opportunities for Mexicans.

In the end, this plan was disastrous for the states and the Mexican people. The vision of Juárez and his administration was to share the profits with the states by using the funds to develop local economies. This plan was undermined after Juárez died on July 19, 1872 (Scholes 1957: 176). The new administration had a different attitude toward the masses, preferring to invest in cities, where elites resided. Using the international infrastructure developed during Juárez's administration, those who took control of the federal government established a class structure in which the upper class depended on US foreign investments to maintain their privileged lives. The modernization of Mexico did advance in the cities, but at the cost of allowing US corporations to exploit Mexican resources.

THE EXPANSION OF US CAPITALISM IN MEXICO: THE PORFIRIAN ADMINISTRATION

In 1876 Porfirio Díaz assumed the presidency in Mexico and entrenched Mexico in a capitalist venture dependent on US foreign investments (Miller 1985: 257). As Mexican capitalism matured, the wealthiest sectors of Mexican society prospered while the common person became entrapped in a production process that benefited foreign investors more than the country as a whole.

President Díaz devised legislation that ran contrary to the social welfare of the masses and the spirit of the Juárez regime. His labor policies betrayed Júarez's vision that a fair compromise between labor and capital must protect the interests of both classes. To encourage economic growth, Díaz devised legislation to encourage foreign capitalists to invest in Mexico. Under his modernization program the presidential cabinet planned to lure capitalist investors by granting them government subsidies, giving them land grants, and awarding them generous tax breaks. Díaz envisioned that his plan would lead industrialists to expand factory employment, bring needed foreign capital into the country, and convince large-scale agribusinesses to increase food production. At that time, legislation in the oil industry was not enacted, as US corporations did not consider drilling for oil in Mexico a profitable venture.

To enforce his modernization program, Díaz created the legal infrastructure to gain control of the country's military and federal police. Once he had secured control of the military and police bureaucracy, he began to impose policies to destroy his enemies and reward his loyal allies. Díaz ordered that anyone who opposed his regime would be imprisoned, sent into exile, placed in work gangs, or shot. To ensure the cooperation of the masses in the rural areas, where three-quarters of the population resided, Díaz increased the size of the federal police force and armed them with weapons that could easily overpower the local police and state militias. The mission of the federal unit, whose members were called the *rurales*, was to ensure that Díaz's economic plan was followed.

As Díaz's power increased, the Mexican Congress acceded to his demands and revised the nation's Constitution to eliminate presidential term limits. This allowed Díaz to be continuously reelected and remain the constitutional president. Díaz and his supporters essentially created a legal dictatorship that was protected by a well-paid military and federal police. To deter dissent, Díaz imprisoned, exiled, or assassinated anyone who questioned his administration's actions. In attempting to develop Mexican capitalism, Díaz chose to eliminate the state's responsibility toward the common person, and he created a cultural hierarchy that made the wealthy the only citizens with legitimate rights and claims to the wealth of the state. This philosophy was influenced by social Darwinist tenets, which proposed that poverty and wealth were the outcomes of the cultural superiority or inferiority of individuals rather than the effects of political practices. Díaz's presidential cabinet, which acknowledged the acceptance of this principle, referred to themselves as the "Científicos." Their administrative policies were allegedly based on scientific facts that would propel the nation toward success, rather than on humanistic policies that only stagnated Mexico's development. According to the Científicos, Mexico's resources must be owned by the wealthy, as only they had the knowledge and capacity to produce capital.

When Díaz took office, in 1876, he announced that his administration was launching a large-scale infrastructure project to modernize Mexico's transportation and road system (Callahan 1932; Lorey 1999). The actual aim, however, was to facilitate the movement of silver and gold to the United States, and to transport luxury goods from the United States to Mexico. Most people did not benefit from the infrastructure improvements, since most towns and cities did not have access to the new railroad routes. Track was laid to connect the mines to the US border and to Mexico City and Guadalajara, where there was high demand for US manufactured goods.

Six US companies were invited to enter joint-investment ventures with the Mexican government and lay railroad track throughout Mexico. The problem was that as roads and track were laid, thousands of families were displaced and their land taken away. Although some property had to be expropriated through eminent domain because it was in the path of the railroad routes, land was also taken to make partial payments to the railroad companies. Díaz paid the companies US$8,000 per kilometer of constructed track and gave them land grants and tax exemptions to encourage them to build more routes.

Díaz also gave land concessions to wealthy Mexican ranchers by transferring titles through eminent domain, or by merely allowing the *rurales* to evict farmers. After people lost their farms by force or legal eviction, the courts issued new titles (Aguirre Beltrán 1991; Murillo 2008; Orozco 1895). Díaz and his supporters proclaimed that such harsh measures were necessary because the only way to modernize Mexico was to put the land in the hands of those who had the assets to develop it. The majority of the concessions, however, were given to the railroad companies. During the late 1870s, the dislocation was gradual, but within a decade the displacement was massive.

The Land Law Act of 1883 facilitated the transfer of private and public land to US railroad corporations (Meyer, Sherman, and Deeds 2007: 398; Murillo 2008). The law required that all public land be surveyed for the purpose of development. Within a few years of its passage, one-fifth of Mexico's total land mass—68 million acres of rural land—changed into the hands of foreign corporations and Mexico's ruling class (Miller 1985: 272). The main beneficiaries of the Land Law, however, were US railroad corporations, as they were the principal investors in Mexico's public-land-surveying projects (Davids 1976: 183). Under the act, Mexican and foreign corporations were hired to survey public lands, subdivide plots for settlement, and establish a transportation infrastructure. In return, corporations were to receive one-third of the land surveyed and be given the privilege of purchasing additional acreage. Owing to the corrupt surveying system established by the federal government, many Mexicans lost their land when the corporations deceptively included privately owned ranch lands. The most coveted property was that on which an irrigation system had been built or which was located near water. Once the land was surveyed, Mexican farmers had to prove legal ownership in court or lose their ranches. Throughout Mexico, the courts upheld the corporations' surveys over the farmers' claims.

Mining corporations also became heavily involved in surveying Mexico's land after Díaz reformed Mexico's mining laws. His intent was to attract foreign corporations to develop federal lands. On November 2, 1884, the Porfirian government nullified Mexico's subsoil law, which under the "Ordenanzas de Mineria" had decreed since 1783 that all minerals in the subsoil were exclusively owned by the state (Rippy 1972: 5). Under the Mining Law of 1884, the owner of the surface soil also became the owner of the subsoil and did not have to acquire government permission to explore or extract minerals (Pemex 1988: 23). This law gave foreign corporations more incentives to participate in Mexico's land-survey projects, as the law of 1883 gave them title over land, while the law of 1884 gave them ownership of the minerals and oil in the subsoil. Within a few years, mining and oil corporations began investing in Mexico by either purchasing property or participating in Díaz's surveying projects.

EDWARD DOHENY: THE FIRST
SUCCESSFUL US OIL INVESTOR

By 1888, US corporations had invested US$30 million in the mines (Callahan 1932: 508). Independent oil prospectors were also attracted by Mexico's generous mining laws, and in the late nineteenth century they began exploring for oil. Among them was Edward L. Doheny, a mining surveyor, who in 1887 took advantage of the Land Act of 1883, obtaining properties in the states of Tamaulipas, Nuevo Leon, and San Luis Potosí (Davids 1976: 192). Unlike his colleagues, however, Doheny was interested in oil rather than in silver and gold exploration. After his first major oil discovery, in 1901, at El

Bano, west of Tampico near the Gulf Coast, Doheny invested heavily in oil exploration and established the Huasteca Petroleum Company (Brown 1993: 29).[2] He then proceeded to purchase property from the railroads. By 1902, Doheny owned 448,295 acres in Mexico. His success attracted others to explore for oil (Santiago 2006: 210).

Before Doheny, a few Mexican businessmen and the American investor Henry Clay Pierce had unsuccessfully explored for oil. Pierce, who owned the Waters-Pierce Oil Company, chose to concentrate in the sale and refinement of oil rather than continue to lose money in oil exploration. Pierce dominated these aspects of Mexico's oil industry until the outbreak of the Mexican Revolution.

US investors also profited from exploiting the labor of Mexican workers. In the cities, they established hundreds of factories, and workers there were not paid a living wage. American investors focused on the production of textiles, plastics, and soap (González Ramirez 1986). To ensure that investors were content with their labor force, the federal government did not protect Mexican workers. Foreign and domestic employers were allowed to force their workers to labor twelve to fifteen hours a day. The federal government also failed to pass protective laws to guard against hazardous employment and did not require employers to pay workers for overtime or compensate them when injured. Employers could also arbitrarily reduce their workers' pay, and a governmental recourse to challenge the validity of these pay cuts was unavailable. To protect company managers from angry workers, the federal government allowed companies to hire their own police, and if the contracted police were insufficient, the federal government would send troops (Miller 1985; Pletcher 1958).

In the rural areas where the mining and oil drilling camps were located, workers experienced similar conditions of long hours and police surveillance. However, they were paid higher wages due to the hazardous work environment and specialized skills required. High wages attracted many employees to these industries, including workers from the United States. The labor force therefore was ethnically mixed in many of the camps, but the companies practiced a dual-wage labor system (Calderón 1975; González Ramirez 1986). Nearly all executives, scientists, and technicians were US citizens, and their wages reflected their level of education and skill. Common laborers were treated differently, however, because of their race. US workers were paid twice as much and in gold, rather than in pesos like the Mexican workers. Mexicans were also expected to work longer hours. If they lived in the camps, Mexicans were housed in shacks without furniture or walls to protect them from the elements, while Americans were segregated in log cabins that had stoves, beds, doors, and windows. Mexicans were also only treated in the company clinic when seriously injured, while Americans had access to it as needed (Santiago 2006).

THE OIL INDUSTRY AND THE PORFIRIAN ADMINISTRATION

To encourage the expansion of the oil industry in Mexico, Díaz began to pass laws to give investors more land concessions. The Land Law of May 12, 1890, allowed the federal government to identify and expropriate the public lands of cities and towns, which could then be used for development (Pemex 1988: 25). Mexicans, including elites, considered the law unfair and were concerned about how their quality of life would be affected if investors drilled oil wells in their communities. Because three years earlier Díaz had prohibited governors and city councils from intervening in development projects or suing corporations for damages, the Land Law of 1890 was potentially disastrous to local environments (Santiago 2006). Concerned citizens took their complaints to the Mexican Senate and presented different legal arguments. The Mexican Oil Union Workers also opposed the law (Rippy 1972).

As citizens launched complaints and met with their legislative representatives, members of Díaz's regime tried to stop the opposition by introducing congressional legislation to clearly stipulate that foreign companies had the right to explore in any location for which they obtained government permission. The Mineral Law of 1892 was introduced within Congress for this purpose, stipulating that landowners had the right to explore for oil and minerals on their property no matter where it was located. Some attorneys from the Mexican Academy of Legislation and Jurisprudence opposed the passage of the 1892 law and submitted their complaints to Congress. Their main concern was to restrict the locations where companies could drill for oil. In addition, they questioned the legal rights foreign companies held over the oil and minerals they extracted. Based on their understanding of Spanish and Mexican law, they believed that the Mining Law of 1884 had not given foreign companies ownership of the oil and minerals they extracted from the ground; the reforms of 1884 had given proprietors' ownership of the subsoil, but not of what was contained within it. Therefore, the attorneys from the academy concluded that Mexico's oil and minerals belonged to the nation, and not to the foreign investors. If the companies were to continue working in Mexico, they must be taxed (Rippy 1972).

In the end, the Mexican Congress ignored the attorneys' legal arguments and passed the law. The legislators also added another stipulation that protected foreign investors. They decreed that in the case of land grants issued to corporations or private citizens, the government could revoke the concession only if federal taxes were continuously left unpaid. The revocation clause did not apply to oil investors, since Díaz a few years earlier had exempted the oil companies from any taxation dealing with production and exploration. Only an export port fee was levied. In 1881 crude oil could not be taxed, and in 1887 the exemption was expanded to include petroleum and any refined oil product (Pemex 1988: 22, 24).

For foreign investors, the Mineral Law of 1892 was a triumph. Their ownership of Mexican land was irrevocable, as long as they followed Mexican law. Pleased with the congressional action, the US Department of State sent President Díaz a congratulatory letter informing him that US government representatives interpreted this law to be irrevocable and permitting landowners to do whatever they wanted with their minerals and oil (Rippy 1972: 22–23). For concerned citizens seeking to regulate the oil industry, the law was a setback. The only benefit Mexicans acquired under the law of 1892 was congressional assurance that oil companies would be restricted from drilling within city limits. Although opponents of the congressional decree accepted the rule of law, they continued to question whether the right to explore and drill gave foreign companies ownership of the minerals and oil (González Ramirez 1941). For now, this issue could not be raised, since the president had personally intervened and asked Congress to clarify the rights of foreign corporations.

In 1894, President Díaz once again ushered a law through Congress intended to attract foreign investors to Mexico. The Property Denunciation Law of May 26, 1894, allowed the federal government to revoke private property titles and transfer title to corporations if the land was abandoned (Pemex 1988: 27). Federal property that had been used for a project and was later abandoned was also subject to denunciation. One of the most questionable aspects of the law was the right given to corporations to identify which lands were vacant and denounce the owners for abandoning the property. Unfortunately, this opened the door for corporations to sabotage landowners by forcing people off the land through intimidation and then declaring the property abandoned.

COMPETITION IN THE OIL INDUSTRY

With the use of the army and rural police, Díaz maintained a tight hold of the population and callously ignored any criticism. By the late nineteenth century, his regime was spending one-third of the national budget on the military and law enforcement (Meyer, Sherman, and Deeds 2007: 396). As the costs for maintaining loyal law enforcers skyrocketed and the modernization programs needed greater infusions of cash, Díaz turned to US investors. When US banks refused to extend loans, the president had to rethink his strategy and was forced to seek the support of British bankers (Gilly 1994; Sloan 1978). Americans were willing to fund development projects in return for government payments and land concessions, but few were willing to issue credit.

British banks became the main lenders, but they placed strict conditions on Mexico. Before any loans were issued, Díaz had to agree to pay the debts owed by Emperor Maximilian. During the French invasion of Mexico, the emperor borrowed money from European banks to finance the war against Juárez's army. After France was forced to retreat, Maximilian was executed and the debts were left unpaid (Zabludovsky 1998: 154). The total loan amount was over 22 million pounds and would be difficult to pay. Bankers realized that Díaz's regime might not be able to meet the quarterly payments, but when assured by the president that any late payment would include additional concessions of public land, they were satisfied with the generous agreement.

The failure of US investors to issue Mexico loans angered Díaz, and the episode became a turning point in the treatment of Americans, particularly in the development of the Mexican oil industry. To Díaz, their failure to issue credit was an insult and signified that Americans believed they had the power to do what they pleased in Mexico. The president needed to respond, but he had to be diplomatic since US investors were Mexico's principal developers. To encourage competition while also gaining more control over the modernization of Mexico, he privileged loyal investors in the bidding of development projects. At the turn of the century, Díaz gave favored British engineer Weetman Pearson massive land concessions to explore for oil. Pearson was heavily involved in financing high-risk joint projects with the Mexican government. In Mexico City he invested in improving the city's drainage and flood defense system, as well as in introducing electricity in many zones. Pearson also invested in routes of the Tehuantepec Railway, which were difficult and costly to construct.

Pearson brought competition to the oil industry and consequently contributed to its expansion. He competed with Doheny in the exploration and production of oil, and with Henry Clay Pierce in the refinement of oil into petroleum products. Pierce's business, the Waters-Pierce Company, refined and shipped oil for most oilmen. Because Pearson's financial assets were much greater than Doheny's, he was able to explore for oil throughout Mexico's Gulf Coast and establish refinery plants to process the oil. His success brought attention to the industry and attracted American speculators. The number of newcomers is uncertain, however, as they were not required to register.

At the turn of the twentieth century, after Pearson placed President Díaz's son on the board of directors for S. Pearson and Son Ltd., he received one-third of the vacant public lands in the Huasteca region (the states of Tamaulipas, Nuevo Leon, and northern Veracruz), which was an area Edward Doheny had previously singularly explored (Santiago 2006: 66). In 1901 Díaz also opened all federal lands for oil exploration, including lands in environmentally sensitive zones such as lagoons and lakes (Pemex 1988: 28). Investors were protected from local lawsuits, as companies could not be sued for environmental damages.

In the Huasteca region, Doheny was encouraged to continue exploring, but was distrusted by the Díaz regime. Doheny was considered to be a disloyal American who repeatedly refused to take the advice of José Yves Limantour, the secretary of finance. Doheny also unwisely hired Pablo Martinez del Rio as his attorney, when it was common knowledge that he was a rival of the president. At times, Doheny also tried to expedite his land grant concessions or lease agreements by making tentative arrangements with governors and local elites rather than negotiating directly with federal staff. In 1905, Secretary Limantour became concerned with Doheny's lack of respect and had his concessions and investments investigated. Although Doheny was allowed to continue exploring, he learned that there were repercussions to his actions.

After the investigation was concluded, Weetman Pearson invited Doheny to partner in several projects in the Huasteca region. Doheny had recently discovered several large deposits (Brown 1993). With Doheny's geological knowledge and drilling expertise and Pearson's laissez-faire access to government land, both men could realistically monopolize the industry. In the near future this partnership could become immensely profitable, since Díaz had recently opened private land for exploration and Pearson was expected to receive privileged treatment in the access to the new lands. In 1905 the Mexican Congress, with the support of the president, gave oil corporations the power to drill in private land, even over the owners' objections (Pemex 1988: 29). If oil was discovered on private property, oil companies were required only to pay a rental access fee and did not have to share any of the profits. By 1906 Pearson's oil property holdings were twice as large as Doheny's, reaching over nine hundred million acres (Santiago 2006: 68).

By 1909, the government projected that besides Pearson and Doheny's companies, independent speculators as well as six other foreign companies were operating in Mexico. The number of oilmen and oil companies drilling in Mexico was uncertain, however, as the government did not require oilmen to register their projects unless they had received a government grant or applied for a lease (Pemex 1988: 28–35). For the Mexican government there was insufficient cause to regulate the industry at this time, since oil production and exploration could not be taxed. Díaz was aware that many Americans had set up drilling projects along the coast, but he remained unconcerned. The president's primary objective was to expand Mexico's oil exploration infrastructure and identify the geological deposits of oil.

The lack of regulation and the increasing number of disastrous oil spills caused Mexican elites to seek support from the Mexican Congress and demand the industry's regulation (Gilly 1994). Many influential Mexicans were also becoming concerned about the massive amount of acreage given to the oilmen. To them it was nonsensical that the president failed to regulate a growing industry and had no policy to revoke land grants when land was abandoned or oil not found. In 1909, to placate Díaz's critics and prevent organized protest, the Congress rescinded the Property Law of 1894 and ended the denunciation policies, which had given corporations the power to declare people's private property abandoned and subsequently apply to recover that land. Congress concluded that current policies were subject to manipulation. President Díaz did not object, since Congress also passed companion legislation that finally settled a scientific question raised by attorneys from the Mexican Academy of Legislation and Jurisprudence. Estevan Ruiz and other attorneys had presented scientific data to Congress that proved oil was not a solid mineral, and because it had not been specifically named under previous legislation, it was exempt from the subsoil laws (González Ramirez 1941: 172). According to their legal arguments, the Spanish Mining Ordinance of 1783 was still in effect, and it gave the state exclusive ownership over Mexico's oil. Their arguments failed to convince most congressmen. To bypass their scientific finding and make their research irrelevant, Congress passed the Mining Law of 1909, decreeing that oil was included as part of the minerals owned by individuals holding a deed of the subsoil (Pemex 1988: 37).

The Law of 1909 ended all legal arguments against censuring the property rights of the oil investors. Within a couple of years the number of oil companies registered in Mexico increased to eleven, with Royal Dutch and two subsidiaries of Standard Oil establishing large operations throughout Mexico (ibid., 33–36). New wells were drilled along the Texas-Mexico border, the Gulf of Mexico, and in Baja California.

Mexico's ruling class directly profited from the passage of the Law of 1909. Soon after, Weetman Pearson and close friends of President Díaz established "Compañia Mexicana de Petróleo El Aguila" (Mexican Eagle Oil Company) (Brown 1993: 63). Pearson and Son, Ltd., became the parent company of Mexican Eagle, with Pearson retaining majority interest. Pearson made sub-

Clearing a jungle prior to locating an oil well in the state of Veracruz, c. 1913. Eberstadt Collection, di_09706, Dolph Briscoe Center for American History, University of Texas at Austin.

Mexican oil workers at Well 401, Pánuco, Veracruz, c. 1910. William Fletcher Cummins Papers, di_09707, Dolph Briscoe Center for American History, University of Texas at Austin.

Warehouse of the Huasteca Petroleum Company, Tampico, Tamaulipas, c. 1910. Created by Gilliams Service, Eberstadt Collection, e_enr_051, Dolph Briscoe Center for American History, University of Texas at Austin.

Bridge building near the Cero Azul, Tampico Oil Field, c. 1910–1920. Eberstadt Collection, e_enr_054, Dolph Briscoe Center for American History, University of Texas at Austin.

stantial investments in the company, especially developing its refining capacity. At this time there were many oil companies exploring in Mexico, yet the refinery industry lacked competition. Independent oilmen and small-scale companies depended on the Waters-Pierce Company to refine their oil and ship it abroad. With the establishment of Mexican Eagle, Pearson and his Mexican investors ended the Waters-Pierce monopoly.

Mexican elites became increasingly concerned with the president's failure to regulate the industry. Large land concessions were given to oil corporations, with very little received in return. The only tax companies paid was a port duty fee of fifty centavos per ton of oil (ibid., 40). In 1909 Francisco I. Madero publicly challenged the president's reckless policies and commenced a social movement to remove Díaz from office. The complaints of the people were many. The land grant concessions, however, were identified to be the main offense affecting most people throughout Mexico. By 1905, several studies revealed that sixty-nine million acres given to foreign and Mexican investors had not produced the expected results. The land was not cultivated or under any type of development (Gilly 1994: 79; see Orozco 1895). Not only had the land concessions led to Mexico's declining agricultural productivity, but many communities were uprooted from their homes and the economies of towns devastated. As peasants moved to the cities in search of work, many did not find employment and were consequently forced to return home if they did not want to starve. Without land, peasants had to accept local wages or concede to tenancy agreements in which they were required to turn over nearly all of their agricultural profits to their landlords. Mexican economists estimated that the daily agricultural wage had fallen to fifteen centavos in 1910 from thirty-five centavos in 1876 (González Ramirez 1986: 21). Employers did not have any incentive to raise wages, since President Díaz allowed landlords to pay workers any wage. In the factories, mines, and oil camps, workers received higher wages, but they were also exploited. Díaz failed to pass a law limiting the number of hours worked or prohibiting children under age seven to work in the factories. Employers did as they wished and forced employees to work long hours (ibid., 48). To Madero the repression of the people had to end. Mexican elites exploited the peasantry, while foreign corporations mistreated the factory, oil, and mining camp workers.

Madero was an industrialist himself and owned large agricultural estates. He realized Díaz's modernization projects worked for the elite and made Mexico's ruling class very wealthy. It was also obvious that Mexico's economy was growing, with exports increasing at an annual percentage rate of 6.1 (Knight 1986: 23). Likewise, foreign corporations were creating jobs. His vision of Mexico, however, was different from that of the president's regime.

Madero believed that workers should be treated humanely and assured a living wage. In his view, Díaz was turning Mexico's governance over to foreign corporations, and under that system the lifestyle of Mexicans could only degenerate. In 1910, foreign investors owned most of the modernization projects, with Americans being the principal developers. Americans owned 78 percent of the mines, 58 percent of the petroleum industry, 68 percent of the plastic factories, 67 percent of railroad companies, and 72 percent of businesses in Mexico (e.g., textiles and soap factories, hotels, theaters, retail stores) (González Ramirez 1986: 679). Madero was also very concerned with the growth of Mexico's foreign debt and the implications this could have for Mexico's future. In 1909, Mexico owed 81.5 million pounds to French banks, British investors, and US railroad corporations (Gilly 1994: 48; Zabludovsky 1998: 179, 184).

DECOLONIZATION: A CALL TO ARMS AND THE MEXICAN REVOLUTION

In 1906 several events took place that alarmed Mexicans and caused elites to question the sanity of President Díaz's regime. In the state of Sonora, thousands of Yaqui Indians were slaughtered by local landowners with the aid of federal troops; in the Cananea Mines of Sonora, Díaz ordered the execution of strikers who protested against their employers' dual-wage system; and in the textile factories of Veracruz, strikers were shot when they demanded higher wages and the end of child labor. Rumors were also heard throughout Mexico that Díaz intended to sell Baja California to the United States in exchange for extraditing the Flores Magón brothers (González Ramirez 1986: 95).

In 1904, Díaz ordered the exile of Ricardo and Enrique Flores Magón when they refused to end the publication of their newspaper *Regeneración*. They were his worst critics and had inspired Mexicans throughout the country to demand political change and establish democratic clubs. Within the middle and upper classes, Mexicans like Francisco Madero had begun to establish political clubs to discuss the president's policies.

When the Flores Magón brothers arrived in the United States they received assistance from the Western Federation of Miners and the Industrial Workers of the World to relocate to St. Louis and reestablish their newspaper. This incensed the president. He could not stop them from publishing their newspaper in the United States and smuggling it into Mexico. Several times Díaz tried to stop the publication of *Regeneración*, but the First Amendment of the US Constitution allowed the publication of the newspaper on US soil. Díaz then attempted to extradite the brothers, but was unable because they had not broken any immigration law (Menchaca 2011).

In 1906, social protests against Díaz quickly accelerated after the Flores Magón brothers published articles reporting the atrocities committed against the Yaqui Indians of Sonora. That year Díaz ordered the removal of thousands of Yaqui Indians after they refused to abandon their ranches in the fertile desert lands of central Sonora. Allegedly, their land contained silver and they were in the way of the mining companies. Using the pretext that the Yaqui communities had declared war against the government, Díaz sent federal troops to assist Sonora's governor, Rafael Izábal, to remove them. For years, the governor had unsuccessfully attempted to use state troops to remove Yaqui communities from the lands they occupied (Calderón 1975; Spicer 1981).

Through *Regeneración* the public also learned of the atrocities committed against the miners in the Cananea camps. On June 1, 1906, a few months after the Indian removals were completed, Díaz sent federal troops to end the miners' strike in the camps owned by the American Cananea Consolidated Copper Company (González Ramirez 1974: xxiv). The company's mines were located close to the border of Sonora and Arizona in the towns of Cananea, Buenavista, and El Ronquillo. Mexican and Anglo American miners were employed in the camps. When management switched from an hourly wage to a labor contracting system that was projected to reduce wages significantly, the Mexican miners went on strike. The policy change applied only to Mexican workers. Among the strikers' demands was limiting the workday to eight hours and receiving equal pay for equal work. Essentially, they asked that they be treated and paid the same as Anglo American miners. After the strikers defied Governor Izábal's order to return to work, he sent state troops to end the strike and asked for federal assistance. On the third day of the strike, while Izábal waited for the federal troops to arrive, he allowed 275 Arizona Rangers and over 200 American citizens to cross the Arizona border and render aid to the company (Calderón 1975: 97). The next day, on June 4, thousands of Mexican federal troops arrived and overpowered the miners. When the strike ended, nineteen miners and four Anglo American volunteers were dead, and ten Mexicans were seriously injured (González Ramirez 1974: 108–109).

When news of the strike spread throughout Mexico, citizens were incensed. To many this signified that their country had finally become a colony of the United States. Governor Izábal had illegally authorized American troops to enter Mexico, an order that only the president and Congress were empowered to do. The public demanded that the governor be placed under arrest. To calm the people's furor, Díaz held a trial. Although Izábal was found guilty, Díaz dismissed the charges and exonerated him. He also gave amnesty to the Anglo Americans who had been found guilty of killing unarmed Mexican miners, but refused to pardon any of the Mexican miners. Instead, he ordered the convicted miners to be executed by a firing squad.

Díaz's harsh policies backfired. Instead of provoking fear in the masses, his violent behavior caused dissenters to defy federal law throughout Mexico and openly demand his removal. A series of labor strikes ensued soon after. Workers from the textile, railroad, and mining industries coordinated their efforts and organized massive strikes. In Veracruz, Díaz ordered striking textile workers to be shot after they refused to return to work. The strikers were demanding an end to child labor, a reduction of the thirteen-hour workday, an end to the practice of discounting wages for holidays and religious days, and improved access to medical care when injured (González Ramirez 1986: 72). In retaliation, the president ordered dissenters shot and instructed the Centro Industrial Mexicano (the agency that oversaw industrial workers) to raise the mandatory workday for textile workers to fourteen hours.

The president's violent response and callous treatment of the workers caused influential citizens of the bourgeoisie to organize democratic clubs. They feared that a revolution was inevitable if Díaz remained in power. In 1909 Francisco Madero decided to organize the Anti-reelection Party and demand free elections in Mexico. Not only was it necessary to remove Díaz, Madero argued, but it was also time to change the regime at all levels of government. Madero began a nationwide tour to motivate people to demand free elections. In 1910, when the president's term of office was ending, Madero ran for president. Díaz responded by prohibiting any form of public organization that opposed his regime and exiling or imprisoning the leaders of the democratic clubs. However, Madero refused to heed the president's warning and continued with his campaign for office. On the day of the presidential election, Díaz was easily reelected, since most people stayed away from the polls in fear of being arrested.

After the election, Díaz ordered Madero's arrest. Within days of Madero's imprisonment, Díaz was pressured to release him, to prevent an armed revolt from protestors. Madero's allies immediately executed plans to smuggle him into Texas. While in San Antonio, Madero met with other exiled refugees and put into action the Plan de San Luis Potosí, which called upon Mexicans to revolt against the Porfirian dictatorship and initiate armed resistance on November 20, 1910 (Gilly 1994: 81). Throughout the country, Mexicans from different social classes acted on Madero's call to arms. Within six months, most cities in Mexico were under the control of Madero's allies. Emiliano Zapata, a mid-scale farmer and local civic leader, took over most of southern

Francisco I. Madero and his advisers, c. 1911. By D. W. Hoffman, Texas-Mexico Border Photo Postcards, di_06002, Dolph Briscoe Center for American History, University of Texas at Austin.

Mexico. In the northern region, Francisco Villa (Pancho Villa) and Pascual Orozco took control of the northern border-states. Villa was a populist leader of working-class origins, while Orozco was a middle-class rancher and mining entrepreneur. Venustiano Carranza, a wealthy landowner, took control of the gulf region and helped Villa and Orozco seize control of the northern states. On May 20, 1911, Díaz resigned, and Madero triumphantly returned to Mexico City (ibid., 94). A few months later, on October 11, 1911, Madero was officially elected constitutional president of Mexico.

Madero planned to maintain Mexico on its capitalist track, but he favored divesting foreign companies of some of the assets that Díaz had unethically usurped from the people. Madero knew that if he was to continue to do business with US and British companies, it would be necessary to respect their investments. These companies, however, had to be regulated and forced to comply with Mexican law. The main problem Madero faced was how to resolve the absentee landlord problem Díaz had created, as the majority of Mexicans did not own land, while 43 percent of Mexico's land mass was owned by American citizens (Sloan 1978: 285). To return the land Díaz had conceded to foreign companies, Madero would need to expropriate part of their landholdings.

Madero planned to institute agrarian reform and distribute property among the needy, as well as return land unethically expropriated by Díaz. However, he ordered that this be done through an expedited court procedure rather than by executive order. This policy angered many military officers, in particular Zapata and Villa, who at this time were the most powerful generals in Mexico. They demanded that the land be returned immediately to the people, arguing that going through the courts would take too long, as the justice system had to be reformed before land cases could be litigated. It would make no sense to litigate cases before removing the judges appointed by Díaz. Madero's policy disillusioned Villa and Zapata, and although they supported Madero, they were not prepared to disarm or cede their regional power to the federal government. Only General Carranza supported Madero's administration afterward. Carranza had been appointed minister of war, a position the other leaders did not think he merited. It was public knowledge that Villa, Zapata, and Orozco, and not Carranza, had placed Madero in power.

THE OIL COMPANIES NEGOTIATE
WITH THE NEW ADMINISTRATION

It is uncertain whether business meetings took place between the oilmen and Madero or Díaz during the Mexican Revolution; if any were held, they were undisclosed. The facts that we do know are only about those events that occurred in public. For example, before the revolution broke out, it was public knowledge that Henry Clay Pierce was incensed with the Díaz regime after the president placed a special export tax on the Tampico Port, where Pierce conducted most of his activities. He was also disgruntled that the Mexican Eagle Company was competing with his export company. Previously, only Pierce's company exported oil for those firms that did not have transport facilities. Edward Doheny was also known to have been displeased with the Díaz regime for favoring Weetman Pearson over the other oilmen. However, although he had several public clashes with Díaz's staff over the land leases, Doheny also received favorable treatment. It was unlikely that he would have supported Díaz's overthrow. He did not publicly support Madero or Díaz, but after Madero took office, Doheny sent a letter to the US State Department asking the federal government not to intervene in the revolution, as siding with the wrong faction could terminate all foreign investments in the oil fields. Weetman Pearson's loyalty to Díaz was public record. We also know that after Madero took office, a well-publicized meeting between Madero and Pearson took place in August 1911. At that meeting, they discussed how Madero's ascent to power would affect the oil industry, as well as the factual basis of a widely circulated rumor alleging that the Standard Oil Company of New Jersey had given Madero US$1 million in exchange for obtaining oil land concessions. Madero assured Pearson that the rumor was false and promised to respect his company's concessions, as long as he followed Mexican law (see Brown 1993: 174–178).

Many, including observers within the US government, believed that the Standard Oil rumor was true. In 1913 the US Senate held hearings in Washington, DC, to investigate whether Doheny and Pierce had instigated war against President Díaz in an effort to obtain some of Pearson's oil concessions. The managers for Standard Oil of New Jersey and the Sinclair Company were also investigated. At the hearing, agents from the US Attorney General's Office testified that in April 1911 representatives of Standard Oil met with Madero allies in El Paso, Texas, to draw oil concession contracts. This took place before Díaz resigned from office. All parties denied any involvement, and with no substantive evidence presented, the accusations against the oilmen were dismissed.

One of the first revolutionary actions taken by Madero in 1912 was to create a new tax base. Taxing foreign companies for their land, as well as for imports and exports, was to become the government's main source of revenue. Most taxes, however, were projected to come from the oil companies, as well as from the small- to mid-scale foreign businesses. Few if any taxes could be levied from the large mining and railroad corporations, since most mines had been abandoned and were no longer functional, while the railroad companies were expropriated by the state and could not be taxed.

In the case of the oil industry, most oilmen had remained in Mexico and continued to drill. The Mexican government estimated that at least five hundred oil companies could be taxed. The government's first tasks were to identify the companies, determine their property holdings, and assess the owed taxes. At this time only the holdings of Pearson and Doheny were public knowledge, with Pearson estimated to own 61 percent of the active oil wells (Gilly 1994: 48; Meyer 1972: 25). On June 3, 1912, Madero levied a stamp tax on the oil companies, setting it at 20 centavos per ton of exported oil, plus a 5 percent tax on the value of the property (Meyer 1972: 63). The property tax was charged to the landowner or lessee. Only Pearson and a few companies complied. Doheny and nineteen other oilmen refused to pay taxes. They immediately sought aid from the US State Department, but were turned down because the US government could not intervene when Mexico was at war. Doheny then filed a lawsuit against the government in Tampico's district court (Pemex 1988: 42–44). Representatives of the department advised Doheny to continue with the suit. In the meantime, it was best for the oilmen to try to settle for a reduced tax. The State Department was confident that Doheny had a valid argument. His property had been acquired under the Díaz administration; therefore, his taxes should be based on earlier agreements.

On February 22, 1913, General Victoriano Huerta, a former ally of Porfirio Díaz, assassinated President Madero, and with the support of the US government he briefly ascended to the presidency (Gilly 1994: 120). Following Madero's fall from power, the oilmen anticipated that Díaz's tax policies would be reinstated. To the surprise of the oil companies, however, Huerta did the contrary and more than doubled taxes. It was his only option, since he needed to quickly finance an army to defeat the forces of Madero's allies. Generals Zapata, Villa, and Carranza temporarily joined forces to remove Huerta. The generals did not agree with Madero's policies, but they considered the assassination of a constitutionally elected president a violation of the principles of the Mexican Revolution. They also realized that Huerta, if not removed, would merely reinstate the government of President Díaz. For the oilmen, Huerta's removal was opportune, as the Mexican Senate chose to rescind his oil tax policies.

Following the death of Madero, the Mexican Revolution exploded as the nation divided into different factions. Each state had a different regional general, with only Zapata and Villa holding mass appeal. At this time, either Villa or Zapata could have ascended to the presidency, but both men were reluctant to do so. The oilmen therefore had to live with the inconvenience of the revolution. To their benefit, major battles did not erupt near the oil fields, and as long as they paid taxes, federal troops left the companies alone. The oilmen were also not prepared to leave, since throughout the Gulf Coast region companies were discovering oil. Doheny's second-most successful oil well, No. 7, the Juan Casino, was discovered on the eve of the Mexican Revolution. By September 1910 Doheny's oil production was at its peak, surpassing Mexican Eagle. Unlike Doheny, Pearson was not enthused with his situation in Mexico and was prepared to sell his assets. He chose not to expand further, and as a consequence the production of his wells declined.

In 1914, Carranza ascended to power and took over the presidency (ibid., 197). Within a year he learned that Villa and Zapata were prepared to remove him from office. They opposed his agrarian reform plans, which, like Madero's, called for the courts to adjudicate the distribution of land. This angered Zapata and Villa, who thought Carranza's plan would benefit only the bourgeoisie and foreign interests. The generals threatened to attack Carranza's forces unless he immediately distributed land through executive order. In the regions controlled by Villa and Zapata, agrarian reform had already been implemented. Carranza feared that unless he disarmed the generals, his administration would not last.

For the US government, Carranza in power was preferable to Villa or Zapata, since the populist generals' agrarian reform plans would likely terminate American property titles. Under Carranza's regime the US government would be able to negotiate some type of restitution for confiscated property, or possibly retain some percentage of the land grants issued to US investors. US president Woodrow Wilson thus officially recognized Carranza's presidency, giving his administration political stability (Coerver and Hall 1984).

Although Carranza's position toward the United States was diplomatic, he resented the economic gains American businessmen had unfairly obtained in Mexico. For every dollar they had invested, they had received tenfold, if not more. Carranza knew that in matters of foreign policy his reforms had to be constitutionally sound and follow international law in order to avoid US intervention and possibly even military intervention. Carranza's immediate concern was to assure the US government that Mexico planned to resume foreign trade and would also continue to allow Americans to invest in Mexico. However, Carranza also had to firmly inform the US government that when Americans were invited to return to Mexico, corporations must obey the rule of law and not expect to be governed under the former dictatorship's policies. In the case of the oil companies, which were not forced to leave or stop drilling during the Mexican Revolution, Carranza developed a new plan to regulate and tax the industry.

On September 15, 1914, Carranza ordered the oilmen, who had not paid taxes or applied for permits, to stop production (Pemex 1988: 53). He also informed the oil companies that their concessions were under investigation. Once their properties were surveyed, government agents would assess a more accurate tax to be paid by each company. The oilmen asked the US Department for advice, as many of them had begun drilling without oil leases after Díaz was removed from office (Rippy 1972). The US government advised them to pay their taxes because Carranza merely wanted to collect revenues and not stop production. The oilmen were assured that the United States was carefully observing the direction the revolution was taking and was prepared to intervene if the oil fields were damaged. Navy ships were under alert and stationed near the coastline of Mexico. By December, most companies complied, and a crisis was averted.

A few months later, Carranza ordered the companies to register, obtain permits, and submit their exploration and drilling plans (Pemex 1988: 56). The Carranza administration needed this information to determine the new tax regime, which was expected to affect production, exports, and the value of property. The oilmen were also informed that Carranza was prepared to recover all oil lands that were inactive or had been abandoned. This initiative was part of Carranza's larger land project to evaluate what percentage of Mexico's land was titled and had valid leases. This alarmed the oilmen, as they feared Carranza planned to nationalize the oil industry.

To increase federal revenue, Carranza's next project was to revitalize the mining industry, as most mines had been abandoned during the revolution. Workers had either left to join the Villistas or destroyed the mines in retaliation for the crimes they perceived to have been committed against them. Most mines had been burned by Villa's troops, forcing Americans to leave. Consequently, production of gold fell by 80 percent and silver by 65 percent (Meyer, Sherman, and Deeds 2007: 475). When news spread that Carranza invited American investors to reopen the mines, Pancho Villa was infuriated. Villa felt betrayed and insulted, as Carranza knew that most mines were located in his territory and that he had ordered their closure.

When Villa received news in January 1916 that businessmen and engineers from the American Smelting and Refining Company were traveling by train to inspect mines in northern Mexico, he ordered an ambush. A group of Villistas captured the train near the border of Texas and Chihuahua by the town of San Isabel. All but one of the eighteen passengers was assassinated. When Carranza heard of the attack, he sent troops to capture and arrest Villa. Villa was prepared for retaliation. He was informed, however, that Carranza had set a trap, using the mining incident as a ploy to create the conditions for American troops to enter Mexico and capture the president's main enemy.

After Carranza failed to arrest Villa, on March 15, 1916, President Wilson sent US troops to capture Villa (Coerver and Hall 1984: 119). Wilson ordered 4,500 US troops, led by general John Pershing, to enter Mexico and arrest Villa. Although Carranza never officially agreed to allow US troops into Mexico, the Villa-Pershing Affair became a national scandal that destroyed Carranza's popularity. The incident suggested that Carranza was a friend of American imperialists. When Villa was not captured, Carranza knew his presidency was in trouble. The only means of changing public opinion was to prove that he was against US imperialism. To do so, he supported the passage of legal reforms to regulate foreign businesses and redistribute land among the peasantry. His most radical reform targeted the oil industry.

AMERICAN OIL COMPANIES ARGUE THAT MEXICANS
CANNOT REVISE THE MEXICAN CONSTITUTION

On May 1, 1917, the oilmen's fears became a reality. The Mexican Congress revised the nation's Constitution, empowering the federal government under Article 27 to exercise its power of eminent domain and claim any property needed for the public benefit of the state (Meyer 1972: 109–112).[3]

(Left) Pancho Villa, c. 1915. Robert Runyon Photograph Collection, RUN00191, Dolph Briscoe Center for American History, University of Texas at Austin.

(Below) Villistas, 1915. Robert Runyon Photograph Collection, RUN00092, Dolph Briscoe Center for American History, University of Texas at Austin.

A legal framework was set in motion to identify property that had been improperly acquired, had been abandoned, or was not in use. The land was to be used for agrarian reform and distributed among landless Mexicans. The denunciation process applied to Mexican and foreign private property alike. Article 27 also nullified Mexico's property and mining laws of 1884, 1892, and 1909. For the oil industry, section 4 of Article 27 was devastating, as it rescinded their ownership of the oil and minerals in the subsoil. Mexico's minerals and hydrocarbons, including oil, were declared the property of the nation.

Following the constitutional reforms, Carranza informed the companies that they must pay more taxes if they wished to continue drilling. A sliding-scale tax was placed on the value of the oil, with a 10 percent minimum imposed. The tax fee would be based on the amount of extracted oil and the size of the property holding (González Ramírez 1941: 58). The 5 percent land tax previously set by Madero's administration was also retained (Brown 1993: 231). Although the oilmen were relieved that Carranza did not impose a 50 percent tax, as was rumored, they still protested and refused to pay. To protect their assets, the oilmen established the "Association of Petroleum Producers in Mexico" (Asociación de Productores de Petróleo en Mexico, APPM). Their main spokesman was Doheny, as well as representatives from Standard Oil New Jersey and J. P. Morgan Banking. They immediately drafted a letter to President Carranza demanding that Article 27 be rescinded; if it was not, they were prepared to litigate its constitutional validity.

The association also petitioned President Wilson to demand that Mexico rescind Article 27. They asked the US government to use force against Carranza and, if he refused to concede, to use the military to confiscate the oil fields. Wilson, in support of the American oil companies, declared Article 27 immoral and illegal under US law. He conceded that under international law, any country could nationalize and expropriate property within its borders, yet he questioned the legality of applying a law retroactively. Although he reassured the oilmen's association that the US government would pursue the matter as a legal issue, he was not prepared to intervene militarily; the United States was on the brink of entering the First World War, and fighting two wars at the same time was unwise. Carranza knew as much and instead waited for diplomatic action. Wilson's secretary of state, Theodore Roosevelt, began legal action to nullify or clarify Article 27. On behalf of the oil association, the US government petitioned the Mexican Supreme Court to determine if Article 27 was constitutional, and to clarify whether it could be applied retroactively. In the meantime, Doheny and other US companies refused to pay taxes and threatened to close the oil fields and fire the oil workers. When Carranza's administration failed to fall to pressure, Doheny threatened Carranza with an economic embargo.

The association asked American banks to deny Mexico credit. When the banks complied, Carranza finally conceded to the association because the Mexican government needed its support. Without credit or tax revenues, Mexico could not survive, nor could its government continue financing the armed struggle against Villa and Zapata (Rippy 1972: 121). Carranza agreed to uphold the Mexican Supreme Court's decision if it ruled on behalf of the oilmen. He also agreed to reconsider his tax regime.

On August 12, 1917, the Mexican Supreme Court offered its interpretation of Article 27. It ruled that Mexico had the legal right to denounce any property for the use of the state, as well as to claim the subsoil rights for the good of the nation. The court also upheld Mexico's right to be the sole owner of its minerals and hydrocarbons. However, the justices also ruled that in the case of the oil industry, the law could not be applied retroactively. Companies that had purchased or were given land concessions before May 1, 1917, owned the subsoil and the oil, as long as the property contained working wells (ibid., 31; Vásquez and Meyer 1985: 130). The court made it clear that this retroactive exemption applied only to the oil industry and not to land concessions purchased or given to mining or railroad entrepreneurs. As part of the ruling, the justices rendered a decision on the oil leases, since the oilmen had challenged Carranza's right to terminate past agreements. On this issue, the court also ruled that as long as oil leases had been issued before May 1, 1917, they could not be terminated. This was a legal triumph for the oil industry, since 90 percent of the land producing oil had been granted or leased to foreign corporations before that date (Meyer 1972: 113).

For the association, this was only a partial victory, because Article 27 had not been overturned. The court's decision presented them with two main problems: first, they could lose the land where working wells had not been established, and second, many of the oilmen could lose the property under their possession. On the latter issue, because many of the oilmen did not have permits, their sites were subject to expropriation without indemnification. After the Mexican Revolution started, many Americans had entered as wild-cat drillers without bothering to obtain permits or pay taxes.

As head of the association, Doheny now turned to fighting all forms of taxation. He asked the US government to demand that President Díaz's tax laws be reinstated. On this point, the US State Department did not support Doheny, since international law did not exempt any foreign corporation from complying with a nation's tax system (González Ramírez 1941). Weetman Pearson, who had supported the association, found this argument illogical and distanced himself from the other oilmen. He was critical of the

American oilmen, seeing them as unscrupulous and disrespectful of the government. Pearson's intent was to leave and sell his assets. In 1917 Doheny and other American oilmen produced 65 percent of Mexican oil (Meyer 1972: 25).

On August 20, 1918, Carranza handed the oilmen good news: he compromised and reinterpreted the Supreme Court's ruling of Article 27 in their favor. Although the justices had limited Article 27 to exempt only certain lands with working wells, Carranza extended the exemption clause to include inactive land that was part of a long-term plan (Pemex 1988: 89). This was a major victory for Doheny, who was one of the few oilmen to hold valid titles and leases over vast undeveloped acreage. Carranza's decree prohibited the confiscation of Doheny's property and prevented other oilmen from setting claim to it.

THE MEXICAN REVOLUTION ENDS

Mexican intellectuals and members of the press felt betrayed by Carranza's interpretation of the Mexican Supreme Court's ruling. In newspapers, Carranza's administration was characterized as a puppet of foreign imperialism for having failed to protect the oil fields. Worst of all, he was denounced for failing to institute his own decree for agrarian reform. Less than 5 percent of Mexicans received land grants during Carranza's administration (Gilly 1994: 260). Intellectuals debated why Carranza had lobbied to pass Article 27 and afterward refused to enforce it.

Nearly three years later, Carranza was assassinated, and within months the Mexican Revolution ended. By April 1919, most people were tired of the fighting, and their constituencies had either put down their arms or fragmented into ineffective civilian militias. Some scholars claim that the revolution ended when Zapata was assassinated, on April 10, 1919, while others put it a few months later, when Pancho Villa retired, and still others when Carranza was assassinated, on May 21, 1920 (Miller 1985: 308).

Many considered Carranza's administration a failure, but others believed he brought stability to Mexico by raising tax revenues and initiating the legal framework to identify, assess, and denounce property illegally confiscated by American and British corporations. Under Carranza's regime, except in the oil fields, foreigners had to prove that they had lawfully acquired their property in order to retain title to it. They were also obligated to prove that their landholdings had never been abandoned or left undeveloped. If all of these stipulations could not be met, their property could be confiscated by the state.

In the case of the oil fields, although companies kept 90 percent of the land with active wells, they lost two-thirds of the acreage. In 1917 they claimed ownership of more than 18 million hectares of land, yet within ten years the Mexican government gradually lowered the estimate to around 8 million hectares (Meyer 1972: 39). Carranza had conceded to the oil companies, but he had initiated a denunciation process to dispossess foreigners of part of their claims. This empowered succeeding administrations to design similar policies.

Carranza's hard-line policies also led to improved working conditions in the oil camps. Although the dual-wage labor system could not be dismantled and companies continued to pay American laborers double or triple the wages of Mexicans, Mexican oil workers received the highest wages in Mexico, and the Mexican government required that their workday be limited to eight hours. In Doheny's and Pearson's camps, living conditions also improved. The Mexican workers' quarters remained segregated, but their shacks were replaced with modest homes and their families were permitted to live in the camps. Furthermore, when the First World War began, Doheny and Pearson gave Mexican workers opportunities for higher-skilled jobs. When Americans left to join the war effort, Mexicans were trained to become skilled and semiskilled mechanics. Upon the return of American veterans to Mexico, Carranza's administration discouraged American companies from replacing the Mexican mechanics, and as a result the number of higher-skilled jobs steadily increased (see Brown 1993). Although life for Mexicans in the oil camps was not ideal, their living and working conditions had improved immensely. The federal government also guaranteed workers the right to form unions for their self-defense, and gradually the workers formed strong unions to challenge their bosses.

Under Carranza, Mexican capitalism entered a new phase. The president introduced the legal framework to regulate foreign corporations. For Álvaro Obregón, the next constitutionally elected president, the challenge was to convince the US government not to interfere with the new regulations. During the First World War, the US government's wartime involvement had temporarily prevented President Wilson from dedicating his full attention to the oil crisis, and by default Mexico's constitutional reforms were accepted. Once the US government was no longer at war, however, the new president had the time and finances to encourage Mexico to reconsider its oil reforms and possibly end the expropriation of American land concessions.

US OIL CONSUMPTION AND ITS IMPACT
ON THE MEXICAN OIL INDUSTRY

In 1920, at the end of the Mexican Revolution, American and British oil companies converted Mexico into the second-largest producer of oil in the world. Mexico produced 90 million barrels of oil that year, which was more than double Russia's production, but less than a third of the US annual production of 360 million barrels (Garfield, Gilbert, and Pogue 1921: 92). Mexican oil was also becoming an increasingly important commodity in the United States, with US geologists projecting that Mexican oil was needed to meet future demand. At that time, coal continued to be the main energy source in the United States for residential and industry use, yet demand for petroleum products was expanding at a rate that required the US government to seek foreign supplies. From 1890 to 1899, Americans consumed 11.8 million tons of crude oil; from 1910 to 1919, consumption exploded to 62.6 million tons (Shao 1956: 82). Coincidentally, US demand for crude oil had increased at the same time that American companies made large discoveries along the Gulf coast of Mexico, with Doheny's No. 7 well at the Juan Casino deposit being the largest oil find.

Economists estimated that demand for crude oil in the United States had increased due to technological innovations in the refinement of oil into energy fuels. Americans had turned to petroleum products to fuel their automobiles, tractors, trucks, ships, machinery, and home lighting. Gasoline for use by automobiles became the most important consumption demand for refined petroleum products. In 1921, 9 million automobiles were in use, and the annual growth was projected to increase by over 2.24 million motor vehicles (Garfield, Gilbert, and Pogue 1921: 248). Consequently, demand for gasoline was expected to more than double every five years. In fact, Americans used 40 million barrels of gasoline in 1915 and 110 million by 1920. If demand for gasoline continued to grow at that rate, geologists projected that US crude oil reserves (i.e., natural geological deposits and stored oil) would be insufficient to fulfill demand. At this rate of use, 90 percent of US oil reserves would be depleted by 1980, unless Americans began importing from abroad. Geologists recommended that the US government begin conserving its oil for future generations, meeting the present demand with exported oil. Two likely sources were Mexican and Venezuelan oil. It was also advisable for Mexico's production to reach its maximum level and for companies to invest in the exploration of light oil, due to its higher quality. Gasoline refined from light oil extended an automobile's mileage per gallon of gas. Although most of Mexico's crude oil was heavy, geologists proposed that its light oil industry could be easily developed. Mexico was also identified as the preferred export nation due to the ease and relatively low expense of transporting crude oil to the United States via trucks or tanker ships.

Thus, following the closure of the Mexican Revolution, crude oil became an important source of energy in the United States, and ensuring its continuation became an issue the US government needed to revisit for the good of the nation. For the US government, it was optimal for Article 27 to be overturned or not enforced, as this would allow American oilmen to control the direction of Mexico's oil industry. America's dominance over Mexican oil became much more likely after Weetman Pearson sold his assets in Mexican Eagle to Standard Oil of New Jersey and the Royal Dutch Shell Company in 1919 (Meyer 1972: 16). Royal Dutch Shell was a binational Dutch and American company. Once Pearson left, Doheny became the most influential oil entrepreneur in Mexico. He expanded his oil investments in Mexico and consolidated his Mexican and US oil companies into the Pan American Corporation. With Doheny leading the way in the 1920s, Americans produced over 73 percent of Mexico's oil (Brown 1993: 123).

UNITED STATES WARNS MEXICO: GIVE US BACK OUR
OIL OR WE WILL BEGIN DEPORTING MEXICANS

When President Obregón entered office, one of his first acts was to study the Mexican Supreme Court's interpretation as to what land was exempt from Article 27. Disagreeing with the position taken by Carranza, he concluded that only land with working wells was exempt (Vásquez and Meyer 1985). He also set in motion plans for agrarian reform, ordering his staff to denounce property, including oil fields that were undeveloped, and begin court proceedings to expropriate it. The US government considered this a hostile position and refused to recognize Obregón's presidency. When US president Warren Harding entered office in March 1921, he took a forceful position against Mexico, demanding that the expropriations stop. He also insisted on repayment for the economic losses American citizens suffered during the Mexican Revolution. To force Mexico into compliance, Harding threatened Obregón with an economic embargo and the deportation of Mexican immigrants. Obregón's administration was shocked. Its representatives had expected economic reprisals, but never before had an American president ordered the deportation of Mexican immigrants in retaliation for Mexico's noncompliance.

When Obregón failed to heed Harding's threats, the deportation raids began immediately. On April 9, 1921, Mexican immigrants in Texas and California were the first to be deported. Obregón had been warned that anyone who had entered the United States without officially registering in a port of entry could be deported, as could those who had become indigent or were unemployed. Only those who had naturalized would not be affected. To stop the deportations, President Obregón agreed to negotiations over future expropriations of land, but refused to concede that Mexico had to pay reparations for land that had been illegally appropriated by US corporations (*La Prensa*, April 28, 1921, p. 1; *La Prensa*, April 11, 1921, p. 1; *La Prensa*, April 13, 1921, p. 1; *La Prensa*, April 30, 1921, p. 1). He asked for a temporary halt of the deportations while he studied the US demands. When President Harding refused to wait, Obregón instructed the consular offices in the United States to aid the deportees and inform Mexican immigrants of their rights (*La Prensa*, April 11, 1921, p. 1; *La Prensa*, April 14, 1921, p. 1). Mexicans who were naturalized and children of immigrants born in the United States could not be deported.

As the news spread of the deportations, thousands of Mexican immigrants were frightened and chose to leave voluntarily (*La Prensa*, April 25, 1921, p. 1). President Obregón learned that city police and county sheriffs throughout the United States were apprehending Mexicans and placing them in trains that would cross the border, leaving families stranded in isolated towns. Deportees were not given food or finances to reach their relatives' homes in Mexico. By May 1921, after over sixty thousand Mexicans were deported, President Harding began the second phase of his plan: the economic embargo (*Dallas Morning Star*, May 10, 1921, p. 1; *La Prensa*, April 20, 1921, p. 1; *La Prensa*, May 4, 1921, p. 1). Neither American corporations nor banks were to conduct business with Mexico. At that point, President Obregón conceded. On May 27, 1921, he notified Harding that he would ask the Mexican Supreme Court to determine whether Article 27 permitted foreign companies to sue the Mexican government for property that had been expropriated (*La Prensa*, May 17, 1921, p. 1). If the justices presented a favorable ruling for US citizens, he would not oppose the indemnification of legitimate claims if acquired prior to May 1, 1917. In the meantime, Obregón outlined a series of executive orders that would immediately resolve some of Harding's main concerns. The most important concession was the agreement to establish a binational commission to investigate the validity of American land titles in Mexico. If the court ruled that reparations were necessary to indemnify Americans, Obregón agreed that the commission would then determine the amount (Meyer 1972: 167; Meyer, Sherman, and Deeds 2007: 507). This was a major victory for the United States, because US commissioners could negotiate financial terms if Mexico's financial proposals were unsatisfactory.

In addition, Obregón rescinded several orders that he and other presidents had instituted during the revolution. First, US oil companies were no longer prohibited from exploring in federal lands (Meyer 1972: 104, 177; Pemex 1988: 104). Second, export taxes for oil would be lowered by 50 percent. Third, the president reinstated Carranza's policy that oil lands without working wells would not be confiscated if they were part of a long-term plan. On this issue, although Obregón was pressured to rescind his directive that "working wells" could only be construed as wells that were drilling oil rather than part of a plan, as Carranza had decreed, the president stood firm on the principle that the government could void a permit on the basis of inactivity. Obregón stipulated that all drilling permits were limited to fifty years, and new permits were required after that period. The oil industry disliked this stipulation and was prepared to fight it. Likewise, Obregón fought Harding's representatives on their demand that the Mexican government not regulate property owned by Americans (González Ramirez 1941: 65–66). This demand was clearly imperialistic and would have led Mexico to lose control over its territory.

By early June, President Harding announced that he would call off the deportation raids since the Mexican Supreme Court was expected to issue a favorable verdict. Talks with Mexico were also progressing smoothly, and he was satisfied with Obregón's proposals (*La Prensa*, June 8, 1921, p. 1). Two weeks after Harding's announcement, the US Department of Exterior Relations informed the press that all deportations would cease on July 1, 1921 (*La Prensa*, June 22, 1921, p. 1).

In Mexico, Doheny and the American oilmen felt that Obregón's executive orders were insufficient. They were against the fifty-year drilling permit limit and dissatisfied that a binational commission was established. Doheny did not want any government agent to interfere in the oil industry's affairs by investigating which oil companies had valid leases and land titles. He just wanted Article 27 nullified, and for Mexico to reinstate President Díaz's tax policies.

To place further pressure on the Mexican government to concede, Doheny advised his associates to purchase the bank loans owed by the Mexican government. The federal government was in arrears, and some banks had begun to sell the loans at a loss. When Obregón learned that the oil association was making deals with the banks, he asked the US Treasury to investigate whether this was an illegal practice (Meyer 1972: 189–192). In the meantime, Obregón launched a public campaign in the United States against the unscrupulous actions of the oil companies. Obregón also contacted the banks and tried to stop them from selling their loans. He reassured the banks that Mexico could repay the loans, and it was best to get full value rather than sell at a loss. The banks agreed to renegotiate, and a crisis was averted. Obregón had won the last battle. The oil companies would not control Mexico's economy, and they would continue to be regulated.

In September 1921, Mexico's Supreme Court rendered its ruling on US reparations (Vázquez and Meyer 1985: 130). The justices upheld Article 27 of the Mexican Constitution, but sided with the US government and gave foreign companies the legal right to sue for their confiscated property. The court ruled that foreigners must be compensated if they held a valid claim. Seeking to strengthen his presidency and resume international trade and free immigration, Obregón endorsed the ruling.

Although agreements made by President Obregón clearly benefited US interests, the Mexican government did not act with deliberate speed to establish the legal framework to institute reparations. The government continued to investigate land claims and stalled all reparation claims. This gave Mexican agents time to identify invalid land claims and denounce the property for having been illegally appropriated.

The US-Mexico General Lands Commission was finally established in September 1923 (Rippy 1972: 90). Initially the US commissioners estimated that Mexico owed US$508,830,321 to American investors for the confiscated property and corporate losses associated with the Mexican Revolution and its constitutional reforms (Sloan 1978: 296; Vázquez and Meyer 1985: 130). Mexico would need to adhere to an annual payment plan of US$30 million for the next four years, with increasing annual amounts in succeeding years. In 1923, the annual payment owed was one-fourth of the federal government's budget. Similar estimates were also calculated for Spain, Germany, Great Britain, and Italy. Although Mexico conceded on the tentative amount owed, it did not agree to pay the money outright. The claims would be separated into different categories (e.g., agrarian, railroad, mining, oil) and a schedule of payments set after the investigations were completed. The final phase was expected to take several years, with no specific date set. Once the commissioners concluded the investigations, a final date for paying the entire amount owed would be determined (Rippy 1972: 94).

In the interim period, the oil corporations continued drilling. It appeared that President Obregón had negotiated a truce. With stability returning to the oilfields, foreign oil companies invested US$862 million in 1923, with 58 percent coming from US corporations (Meyer 1972: 25). Seven new refineries were built to process more of the oil that was gushing from wells throughout the Veracruz and Tamaulipas Gulf Coast (Santiago 2006: 112).

Under Obregón, little advancements were made to regulate the oil corporations. Mexico had taken a step backward and conceded to US demands. The US government had resumed its imperialist stance and demanded reparations for land that had been questionably appropriated. It had also taken the position that Mexican oil belonged to Americans. Mexico was being invaded again, not by military forces as in 1848, but via diplomatic policy and economic agreements. It was a new political era for Mexico. The US government exerted undue influence over Mexican economic policy because after the Mexican Revolution, the federal government needed US dollars to finance the country's reconstruction. The Mexican government had to follow US rule if it was to obtain financial assistance and credit from the United States. During Obregón's administration, the Mexican Congress had to choose whether to continue a pattern of economic dependency, or rupture this reliance and live on its own finances. Congress chose the capitalist route—to borrow—and made concessions.

A DEFENSIVE STRATEGY: MEXICO'S ALIEN LAND LAWS

Federal policies toward the oilmen remained favorable until Plutarco Elías Calles took office. President Calles disagreed with the oil agreement Obregón had negotiated with the United States, and within one year of ascending to office he obtained the support of the Mexican Congress to regulate the property owned and leased by the oil companies. On December 26, 1925, under the *Ley Reglamentaria de Articulo 27*, the Mexican Congress decreed that property owned by oil companies that remained inactive for fifty years returned to the state (Meyer 1972: 252). If companies planned to retain possession of any inactive property, the land must be surveyed and an application filed to retain ownership. After reviewing the application and a company's long-term plans for the inactive land, the government would then determine whether a permit giving the oil company control of the land for fifty years was warranted. Depending on the government agent's determination, a fifty-year permit would be issued or denied. Lands with active wells were protected by the Mexican Supreme Court's ruling of 1917, and no limit on their ownership could be set. However, all properties, whether active or inactive, required a survey and permit.

As part of the new regulatory petroleum laws, Congress also decreed that January 4, 1927, was the oil companies' final date to submit documents proving that their corporations held valid land titles (Pemex 1988: 122, 125, 129). This deadline also applied to companies holding leases. If the oilmen did not comply, their drilling permits would be immediately terminated. Congress also decreed that companies that had not paid taxes would be prohibited from drilling. Finally, the Mexican Congress decreed that all drilling permits were limited to fifty years and must be renewed.

British corporations complied with Calles's deadline directive, while most American oilmen ignored the warning and waited for the US government to intervene. On April 12, 1926, US president Calvin Coolidge sent a letter to President Calles protesting the new petroleum laws (US Congressional Serial Set 1926: 1). He regretted that the Mexican Congress had broken President Obregón's agreement, and he alleged that the government had failed to uphold the Mexican Supreme Court's ruling stipulating that Article 27 could not be applied retroactively. President Coolidge demanded that the oilmen's rights be reinstated. Mexico's minister of foreign affairs, Aaron Saenz, responded that the Mexican Congress had the right to reform its laws. The Congress was incredulous that any government would expect another government to retain its laws in perpetuity. The main argument invoked by Saenz centered on the international principles governing alien land laws. The minister stated that most nations accept the legal principle that a nation cannot protest against laws affecting its citizens in foreign nations, if that nation practices similar laws. In the case of the United States, the alien property laws of the states were the most stringent in the world. Most states required aliens to be naturalized to own property, and at the death of an alien, a foreigner could not inherit said property. Not only that, Saenz wrote, but some states limited foreigners to owning 30 percent of a US corporation. In the case of the Mexican petroleum industry, Mexico's laws were very generous. Mexico did not require the oilmen to be US citizens or place a percentage limit on the foreign ownership of a petroleum company. Mexico did not compromise at this time.

On January 4, 1927, after many American companies failed to comply, Calles sent troops to a few camps and closed operations. US ambassador Dwight Morrow immediately called Calles and asked him to extend the deadline. Morrow agreed to personally ask the oil companies to comply if the president removed the federal troops. Although the ambassador was conciliatory, he warned Calles that the US government was prepared to fight his policies and had joined the American oilmen in their suit against the government. The American oilmen had once again taken their complaints to the Mexican Supreme Court, expecting a favorable decision.

When Calles refused to rescind his orders, the American oilmen stopped paying taxes, fired half of their workforce, and asked American banks to stop all transactions with Mexico. The confrontation lasted several months and was not resolved until November 17, 1928, when the Mexican Supreme Court rendered a compromise. The court ruled that the Mexican government had the right to require companies to survey their property and apply for permits. The right was based on the power Article 27 bestowed on the nation. The justices stipulated that the oil belonged to the nation, and the federal government had only bestowed upon foreign companies the right to extract oil; therefore, permits were needed. This stipulation clarified the court's interpretation that the nation owned the oil, regardless of whether some oil companies may have owned the subsoil.

Then, on behalf of the oil companies, the court ruled that the Mexican government did not have the right to denounce and expropriate property that was acquired before Article 27 was adopted. This applied to land with active wells or undeveloped property (Meyer 1972: 272; Pemex 1988: 130–133). The government also did not have the right to place a time limit upon the number of years a company chose to drill. Only land that was leased was subject to a fifty-year-limit permit. The ruling protected large-scale companies that had purchased property or were given land grants before the Constitution was revised in 1917, while at the same time forcing those who did not have title to an oil field to comply with Mexico's new regulatory laws.

In this contentious battle, the main victory for Mexico was the ruling that ordered oil companies to submit surveys and prove valid custody. This allowed the Mexican government to determine the amount of land under the possession of the oilmen. By the end of Calles's time as president, Mexican agents determined that the oilmen had vastly reduced the amount of acreage they claimed, limiting it to around 8,222,063 hectares (roughly 20.3 million acres) of leased and titled properties. Whether the leases and deeds were valid, however, would be left for the US-Mexico General Land Commission to determine (Rippy 1972: 170). For now the property could be taxed. Once the oilmen began to comply and survey their property, Calles backed down and suspended the fifty-year-limit permit for leased property.

Without a doubt, the property under the oilmen's possession was vast, and the government's recovery of the concessions given by Díaz seemed an impossible task. Mexican agents, however, continued their diligent review of titles to reduce the oilmen's claims. Throughout Mexico, land agents found that most American property claims were invalid, particularly those issued for agrarian use. During the Calles administration, most agrarian claims were concluded to be fraudulent, left undeveloped, or inactive. These lands were not protected by the oil policies. Agents for the land commission concluded that out of the millions of agrarian hectares under review, only 2 million hectares claimed by Americans appeared to be valid, and only 200,000 hectares (494,000 acres) were under cultivation or used for ranching (Meyer 1972: 225; Rippy 1972: 109).

With valid agrarian claims, however, the Mexican government found other ways to nullify the deeds. Mexico's Alien Land Law of 1925 was enforced by the Calles regime. The law stipulated that Mexican nationals must own 50 percent of the acreage used for agrarian purposes if foreign investors owned the corporation. If foreigners were the sole owners, then 50 percent of the acreage must be sold to Mexican citizens (US Congressional Serial Set 1926: 11). The US government initially protested against Mexico's alien land law, but since US alien property laws were much more stringent, there were few legal arguments to advance in support of Americans. The US government had to be content that this alien land law had not been extended to the oil corporations.

During the late 1920s, the Mexican government was well aware that any land that was expropriated must be indemnified. Therefore, for the welfare of the nation it was best for the denunciation and certification of claims to proceed at a slow pace in order to carefully identify fraudulent claims and, if a claim was valid, delay the indemnification process for as long as possible. This was a setback for American mining and railway entrepreneurs who held valid agrarian claims, since payment could take decades. Their claims were also susceptible to citizen denunciations, since the government allowed Mexicans to identify property that was improperly acquired. The slow pace was also beneficial for American oilmen, especially for those who knowingly held invalid or imperfect concessions, because they could obtain temporary permits and continue drilling with little intrusion from federal agents. In 1928, Edward Doheny sold his oil leases and land titles; most were purchased by Standard Oil (Meyer 1972: 16).

ON THE EVE OF THE GREAT DEPRESSION

In 1927 the price of oil began to fall and the oil companies requested a tax reduction. When the Mexican government declined their request, the companies threatened to move their operations to Venezuela, where the oil industry was unregulated. This time, rather than complying with their threats, the Mexican government informed the companies that if they closed operations, their leases and deeds could be terminated under Article 27. The companies did not move, but they reduced production. In October 1929, after the US stock market crashed, the US economy was engulfed by the Great Depression and demand for Mexican oil dropped. Mexico continued to export 40 percent of its oil to the United States, yet its global demand fell by over 60 percent. In 1929 annual production fell to 44,687,887 million barrels of oil, down from 115,514,700 in 1925 (ibid., 21). When sales plummeted, many US oilmen prepared to close operations, but they were pleasantly surprised when Mexican domestic consumption allowed the industry to survive.

Mexicans were consuming most of the oil produced at home, and this prevented the collapse of the industry.

During the Great Depression, Mexico experienced a sharp drop in its commerce with the United States and a radical reduction of its tax oil revenue. But Mexico endured the economic collapse much better than other Latin American countries, because it had a reserve of 30 million pesos (US$15 million) to maintain its economy. Furthermore, since over 70 percent of the population subsisted on agriculture, most Mexicans were self-sufficient. The main area where Mexicans experienced depressed conditions was in factory employment. This primarily affected cities, where factories employed a significant percentage of the workforce.

Mexicans who had immigrated to the United States also felt the shock of the Great Depression, and their treatment across the border strained US-Mexico relations. The US government began to deport Mexican immigrants on a grand scale. It is uncertain, however, whether the deportations were associated with the events that spurred the Great Depression or were in retaliation against President Calles's regulatory laws. US congressional plans to address the Mexican immigrant problem began two months after the oilmen were ordered to submit documentary evidence that they had complied with Mexican law. In March 1927, Congress began holding public hearings to discuss reducing immigration from the Western Hemisphere (*Immigration from Countries of the Western Hemisphere: Hearings before the Committee on Immigration and Naturalization, House of Representatives* 1927). The hearings directly targeted Mexico, since most bills asked for a 90 percent reduction of Mexican immigration, while Canada's numbers were virtually left alone.

At this time, Congress chose to leave immigration alone, and the dispute between the Mexican government and the oilmen ended. Relations actually improved, and it appears that during the Great Depression the American oil companies and the Mexican government worked together for their mutual benefit. Americans continued to drill, and Mexicans consumed most of the oil, which in turn provided a steady flow of income to the companies.

It is most likely that the mistreatment of Mexican immigrants was related to the depressed US economy, and not to the oil dispute. When the US economy began to slow down in February 1929, the US Congress resumed its plan to reduce global immigration and commenced talks to deport unwanted aliens (US Congressional Serial Set 1929). From 1929 to 1933, approximately 400,000 Mexican immigrants, including their US-born children, were deported (Ngai 2004: 235).[4] Most deportations occurred in the cities, where Mexicans were accused of taking jobs away from American citizens. To assist the Department of Labor in removing unwanted aliens, Congress raised the department's budget in 1931 by an additional US$500,000. The department was commissioned to identify cities where a reduction of the population was needed, and to allocate funding to local police and county sheriff divisions to carry out the deportation raids (US Congressional Serial Set 1931a: 2). Studies conducted by the House Committee on Immigration and Naturalization had prompted Congress to take action against Mexican immigrants. One of the committee's reports, titled "Temporary Restriction of Immigration," warned Congress of the injurious effects Mexican immigrants posed to American citizens. Mexican immigration was alleged to cause a racial commingling problem in the United States, introducing criminal and social elements that were harmful to the character of the US population. The report stated:

> That study disclosed, and the whole committee found, that every reason which has prompted the United States to restrict mass immigration from Europe and Asia urges restriction of immigration from Mexico. That immigration is creating another serious race problem in the United States; it tends to displace American workers and has added substantially to the seriousness of the present unemployment conditions. It creates grave sanitation and health problems, tends to overload American charities, adds to the difficulty of proper law enforcement, and contributes further to a racial commingling, which threatens serious injury to the character of our population. (US Congressional Serial Set 1931b: 13)

As the US economy recovered, the deportations ended in 1934 and Congress resumed its nonimmigration quota policy toward Mexico. Nonetheless, the cruel and inhumane treatment of Mexican immigrants had strained US-Mexican relations. Deporting Mexicans was a hostile and unfriendly policy to enact against a neighbor with a history of passing trade policies that generated enormous profits for US corporations, in particular the oil industry. Once nearly half of the Mexican immigrant population was estimated to have been deported, the economic balance between both nations was shattered. In 1934 the newly installed president of Mexico, Lázaro Cárdenas, was highly critical of the asymmetrical treatment of foreigners in Mexico and the United States. The US government expected Mexico to respect the property rights of American citizens and give oil companies rights superior to those that Mexicans held, yet it was not willing to do the same for Mexicans (see González Ramirez 1941: 94). Cárdenas believed that unless he addressed this asymmetry, Mexico was doomed to become a semicolony of the United States. It was therefore necessary for the Mexican Congress to revisit the political rights of foreigners on Mexican soil and, if needed, revise Article 5 of the Law of Nationals and Naturalization. Mexico's alien property laws were very generous and perhaps needed to be amended.

The president's review of Mexico's alien laws coincided with political unrest in the oil camps. In 1934 ten thousand Mexican oil workers organized themselves into nineteen unions and demanded higher wages, improved working conditions, increasing employment of Mexican scientists, and placement of Mexicans in management positions. By 1934 the oil industry had made a full recovery. American and British companies were exporting nearly US$1 billion in oil, with 52 percent of the exports owned by Americans. Mexican taxes had also remained stable during the Great Depression and were low in comparison to the taxes paid by the oil companies in the United States. In Mexico, companies paid on the average 1.05 pesos (US$0.29) per barrel, while they paid the US government US$1.14 (Rippy 1972: 175). With this generous tax system the companies certainly could afford to pay the workers better, in particular the subsidiaries of Shell Oil Company, which had recently discovered massive oil deposits in Poza Rica, Veracruz.

After a series of strikes, Cárdenas asked the oil companies to arbitrate and reach a compromise. When the companies refused, he established an arbitration board called the Federal Board of Conciliation and Arbitration (Junta Federal de Conciliacíon y Arbitraje). On December 18, 1937, the federal board presented its findings, which upheld the workers' claims (Meyer 1972: 319). During Cárdenas's administration, labor unions received unprecedented support. Unions were helped to consolidate into a confederated national union with the power to call national strikes. When oil companies refused to comply with the federal board's findings and refused to acknowledge the legal standing of the unions, Cárdenas supported the unions' right to take their claims to court. The oil companies had resisted the board's ruling under two main pretexts: 1) the federal government could not dictate whom the companies allowed to be part of management, and 2) raising their employees' wages was absurd as they were the best-paid workers in the country (Jayne 2001: 31). In 1938, after a series of court hearings, the Mexican Supreme Court upheld the lower court's decision in favor of the unions.

US and British executives refused to obey the Supreme Court's order and in retaliation threatened to withdraw their investments—behavior that the Cárdenas administration considered a flagrant disregard of Mexican law. Ambassador Josephus Daniels advised the American companies to comply with federal law. The wage demands were reasonable and would not adversely affect the companies' profits (ibid., 44). The ambassador also supported placing Mexicans in management positions. Daniels advised the American oilmen that under the new regime, they had much more to lose than gain for not compromising. The US government also did not want this incident to reopen heated national discussions over the interpretation of Article 27 and Mexico's alien land laws, as President Cárdenas, unlike his predecessors, was likely to implement major reforms.

To Cárdenas, the response of the companies was a belligerent act of aggression against the Mexican government (González Ramirez 1941: 44). US-Mexican relations were becoming intolerably unbalanced and were beginning to parallel the position Americans had taken during the Porfirian period. That is, in the last few years, the US government had shown little regard for the welfare of Mexican immigrants. And now US oil corporations refused to obey Mexican law. It was critical for American companies to be forced to comply, as they were the leaders in this standoff. US investors owned 70 percent of the oil fields. To Cárdenas, if foreigners were unwilling to abide by Mexican law, then their property rights must be rescinded (see ibid., 44, 266). Mexico was not prepared to return to the Porfirian days.

On March 18, 1938, President Cárdenas signed a decree nationalizing Mexico's oil industry (Meyer, Sherman, and Deeds 2007: 532). The holdings of all oil companies, including seventeen US oil companies, were expropriated. Foreigners could no longer own property containing oil. The president no longer felt obliged to continue with the oil agreements merely because foreigners had developed the oil industry. Oil companies had recovered their original investments several times over, and it was now time for oil profits to belong to Mexico. The Mexican government estimated that US corporations had earned between US$1 billion and $5 billion in profits from an investment of US$100 million to develop the Mexican oil industry (Meyer 1972: 31).

Mexico received support from throughout Latin America. Cárdenas was praised for moving Mexico toward an independent economy. Germany and Italy also congratulated the president and requested increases in their commerce with Mexico. The German government was especially interested in helping stabilize Mexico's oil refineries to ensure that oil productivity did not fall (ibid., 37). Mexico would not be able to maintain its production levels unless the refineries returned to full capacity. The governments of the United States and most of Europe denounced the decree and advised the oilmen to follow legal action. The US government, however, immediately informed the American oil companies that armed force would not be used, as this would likely push Mexico to accept Germany's support. President Franklin D. Roosevelt did not wish to retaliate, as war loomed between Germany and Europe and it was highly probable that the United States would need to intervene. Becoming hostile toward its neighbor was not a good idea when Germany was trying to convince Mexico to become its main Western Hemispheric ally (ibid.).

Roosevelt also did not want the oil dispute, which he believed to be a temporary conflict, to threaten the land claim reparations the US-Mexico General Land Commission was finalizing on behalf of American firms. In essence, the US government expected Cárdenas to back down within a few months after he felt the financial shock of the expropriation decree.

THE US GOVERNMENT COMPROMISES
FOR NATIONAL SECURITY INTERESTS

US representatives of the US-Mexico General Land Commission were instructed by President Roosevelt to continue discussions over war reparations owed to American citizens. From November 9 to 12, 1938, Mexico and the US agents settled most land claims, nearly all of which belonged to railroad corporations. Of the 2 million hectares considered valid claims, Mexico agreed to indemnify investors for 1,936,726 hectares (Rippy 1972: 109). The United States demanded US$23 million, but Cárdenas agreed to pay only $12 million. Cárdenas had refused to compensate the railroad corporations for land grants that had not included some type of monetary transaction. Most property, Mexican inspectors also argued, had been forcibly taken from Mexican citizens when US corporations drew surveys and knowingly included private property within their boundaries (ibid., 111). The US government accepted the commission's findings, in anticipation that diplomatic relations would be needed if a favorable deal was to be negotiated for the oil companies. US officials hoped that Cárdenas would rescind his decree. Once the expropriation proceedings concluded, Cárdenas enacted the largest agrarian reform program in Mexican history, awarding 49,580,203 hectares to 775,845 landless families (Gilly 1994: 362; Meyer, Sherman, and Deeds 2007: 528). The Mexican Revolution's main goal had finally come to fruition.

After President Cárdenas failed to rescind the expropriation decree, Great Britain declared an economic embargo on Mexico and asked the US government and other nations to support it. The British government also threatened to call in the loans owed by Mexicans. Cárdenas responded by severing international relations with Great Britain, relying on the US government to be Mexico's intermediary. The president planned to indemnify the companies for expropriating their assets, but this had to be done through negotiations and not threats. British banks tried to place pressure on Mexico to concede by filing claims in US and French courts demanding full payment of issued loans. The courts repeatedly ruled against the banks, either on the basis that only Mexican courts could adjudicate these cases or because the contracts as executed permitted British banks to call in the loans only if Mexicans failed to make their scheduled payments.

British reaction to Mexico's oil expropriation was harsh and immediate because, unlike American corporations, they had to litigate their claims in Mexican courts, which they considered a futile act. Under Mexico's alien land laws, most British investors had chosen to incorporate their companies as Mexican national companies so that they would be exempt from foreign investment policies. This incorporation now prevented British investors from using international law in their defense, since the companies were theoretically Mexican. In contrast, American companies were foreign companies operating on Mexican soil and thus were subject to full compensation. Without legal recourse under international law, British firms chose to ruin Mexico's economy by enacting an economic embargo, as they perceived this to be the only approach that might reverse Cárdenas's expropriation decree.

At first the US government did not support Great Britain's position. However, within a year, when it became obvious that Mexico was not prepared to return the oil fields, the US government unofficially joined the boycott. President Roosevelt instructed the US State Department to prohibit US firms from selling oil-drilling machinery to Mexico. To keep Mexico's oil industry functioning, Cárdenas established a state-owned oil company called Petróleos Mexicanos (Pemex), giving it a national monopoly over oil exploration, production, and exports.[5] Only Pemex was authorized to handle all aspects of Mexico's oil industry.

The US Department of State was also authorized to veto any contract to sell Mexico rayon fabric, electrical material, oil machinery parts, and food. To damage Mexico's economy, the US Treasury was instructed to reduce commerce by boycotting Mexico's main revenue trade sources. Mexico's silver imports were reduced in half, and high tariffs were placed on Mexican oil to make it uncompetitive in American markets. US banks were also discouraged from lending money to Mexico. The US State Department also asked Latin American countries not to conduct business with Mexico.

The American oil companies retaliated by officially joining the boycott. Their first act was to withdraw millions of dollars from Mexican banks to destabilize the economy. Some of the companies also took a violent route, trying to destroy Pemex by sabotaging the oil fields. Before American workers left, they broke machinery, shredded all scientific reports, and destroyed geological data. They also confiscated ships, tanks, and vehicles that the Mexican government had expropriated. Since this did not stop Pemex from producing oil, American companies paid Mexicans to burn oil wells and destroy machinery (Meyer 1972; Pemex 1958, 1988). Their intent was to damage machinery and then call a global embargo on the sale of parts needed to make repairs.

The embargo was effective, but it did not stop Mexico from producing and selling its oil. Mexico began to manufacture its own parts, and Germany and Italy increased their orders of Mexican oil (Pemex 1958). Guatemala, Cuba, Nicaragua, and Brazil also broke the embargo. Germany and Italy alone consumed 65 percent of Mexico's oil in 1939, and in exchange traded needed basic commodities (Jayne 2001: 91). With the embargo triggering an economic depression in Mexico, President Cárdenas presented President Roosevelt with a plan of action mutually beneficial to both nations. Although the nationalization of the oil industry would not be rescinded, Cárdenas agreed to allow US firms to invest in the oil industry. Pemex would manage and hold majority interest in all projects. In return, US firms would be given minority representation in the Pemex board of directors, with at least a two-thirds membership composed of Mexican citizens. Roosevelt felt this was a fair compromise, but when he presented it to the American firms it was immediately rejected. The oil companies responded that they were not willing to negotiate for anything less than full control of the oil industry, plus ownership of the subsoil rights. The unrealistic position chosen by the oil companies angered Roosevelt, because Mexico had taken an irrevocable stance on the subsoil rights. This was a dead issue. The companies needed to negotiate their reentry diplomatically, rather than behave as if they owned the companies and Mexicans had no say in the matter (ibid., 151).

In June 1940 it became necessary for President Roosevelt to bring the oil conflict to an end (ibid., 141). Germany invaded France, and sending military troops to assist Great Britain in combating the invasion was inevitable. Roosevelt immediately contacted President Cárdenas for support and military assistance. Cárdenas agreed to form a joint defense commission, but first required that an agreement be enacted bringing closure to all American claims over expropriated property, including the oil concessions. Roosevelt agreed and commissioned Ambassador Daniels and the US secretary of state, Cordell Hull, to finalize the land claims and defense agreements before Cárdenas left office. Cárdenas's presidency was ending, and for US national security it was best to certify the agreements. It was uncertain what foreign policies the next president would institute. The frontrunners for the Mexican presidency were Juan Andreu Almazán, who in his campaign promised to develop closer relations with Germany, and Manuel Ávila Camacho, a conservative ideologically aligned with the Cárdenas regime. At this point the only issue that the US government knew both candidates supported was upholding the oil nationalization decree.

In December 1940 the Cárdenas-Roosevelt wartime agreement was placed into action and the Mexican-American Joint Defense Commission was established. Members of the Mexican Congress immediately ruptured all relations with Germany and prohibited the sale of oil to the Axis Alliance. As part of the agreement, the Mexican Congress allowed US air bases to be established on Mexican soil and commissioned Mexican pilots to join the US Air Force. In turn, the US government financed a new oil refinery plant to be used for the air bases, and Mexico agreed to sponsor a defense plan for the Panama Canal.

On November 19, 1941, Manuel Ávila Camacho became Mexico's president and upheld Cárdenas's agreements. During his term, the US government finalized the Mexican Revolution land claims and brought closure to the oil claims as well as other financial matters dealing with the war effort. Mexico agreed to pay US investors US$40 million, of which the oil companies were to receive nearly $24 million (Meyer, Sherman, and Deeds 2007: 532; Rippy 1972: 114, 311). A year earlier the Sinclair Oil Corporation had settled for US$6 million, so its claims were not included in the total reparations sum.

Mexico did well, agreeing to pay a much smaller debt than the one Obregón had been pressured to accept in 1923. To that date, Mexican government agents had successfully delayed payment of Obregón's agreement and were able to renegotiate the final sum (González Ramírez 1941: 39). The government's strategy to extend the claim negotiations for nearly two decades had proved wise. In the meantime, Mexican inspectors were able to denounce property, expropriate it, and then litigate a settlement. In the case of the oil companies, after US investigators examined their claims, they concluded that the estimated total sum of US$400 million was exaggerated (Meyer 1972: 451). With the assistance of the US government, Mexico also reached similar agreements with other countries.

To fulfill Mexico's war agreement, President Camacho borrowed US$30 million from the US government. The funds were used to improve Mexico's transportation infrastructure, specifically the construction of Mexican highways connecting the US border to Mexico's southern border. The goal of the project was to facilitate the movement of people and vehicles from the United States to Latin America. The US government also offered Mexico US$20 million in credit for general development and improved communication systems, and agreed to stabilize Mexico's silver market by purchasing 6 million ounces of silver on an annual basis (Jayne 2001: 148; Meyer 1972: 448).

American oil companies felt abandoned by the US government and refused to accept the settlement. The British government, however, accepted the US position and ended its official embargo of Mexico. With the Second

World War escalating, the United Kingdom needed Mexico to be an ally. Although the British government conceded, British and American oil companies refused to do so. This was harmful to the Mexican oil industry, since the oil companies continued the embargo of machinery, technology, and parts needed by Pemex. In spite of the hardship the oil companies forced Mexico to undergo, Pemex continued producing and exporting oil. In 1938, 3.7 million barrels of oil were exported, and by 1947 sales had grown to 6.7 million barrels (Pemex, *Statistical Yearbook 1977*: 30). During the embargo period, however, more than two-thirds of Mexico's production was internally consumed because most American and British companies refused to purchase Mexican oil (Pemex 1958).

In 1943 American oil companies finally settled with Mexico, and British firms settled in 1947 (Meyer 1972: 457). Although the embargo officially ended when companies signed the reparation agreements, Pemex reported that most companies refused to sell machinery and parts until the late 1950s (Pemex 1958). Pemex directors reported that this tactic was taken to force the Mexican government to reconsider a new agreement. Apparently, American companies were now prepared to accept the resolution Cárdenas had offered during the embargo. Unfortunately for them, the Mexican government was no longer interested. Relations between the oil companies and the Mexican government remained strained until the 1970s. Likewise, the US government periodically attempted to use its political clout to pressure Mexico into discussing the reentry of American firms (Brown and Knight 1992; Flores-Macias 2010). It was not until the mid-1970s that the US government stopped pressuring Mexico to negotiate. This conciliatory position was triggered by events taking place in the Middle East. The US government had become embroiled in conflicts with the Organization of the Petroleum Exporting Countries (OPEC), and it quickly learned that for security interests it was best for Pemex to become a successful company.

In retrospect, the administration of President Cárdenas changed the course of Mexican history, with Mexico becoming less dependent on the United States. Many of the goals of the Mexican Revolution were finally instituted. The people obtained some of the land that had been usurped by the Porfirian administration, and property under the control of foreign corporations returned to the nation. The most momentous event was the conversion of the oil industry into a state agency whose profits were to be used for the development of the nation. After the expropriation decree and the end of the economic embargo, the Mexican economy improved significantly. The Mexican government invested in infrastructure improvements, establishing low-cost housing projects, parceling out over thirty million acres, investing in agricultural development, and improving the health condition of the nation.

The Mexican government also invested heavily in schooling, and the literacy level of the nation improved. In 1910, 78 percent of Mexicans were illiterate; by 1960 the number had fallen to 42 percent (Meyer, Sherman, and Deeds 2007: 576). During the 1970s, the poverty level fell and wavered between 25 to 34 percent of the total population (Lustig 1995: 83; World Bank 1980: 4).

The main problem for Mexico, however, continued to be its dependence on US financial assistance. President Camacho's wartime loans left Mexico in tremendous debt and spurred a pattern that became unbreakable in succeeding presidential administrations. Loans could be easily obtained from the United States. That is, the US government and American banks were more than willing to assist Mexico, as long as lucrative interest rates were agreed upon. In the aftermath of the Camacho administration, US business ventures in Mexico were no longer based on taking possession of Mexican land, as was common in 1848 and during the Porfirian administration. US investors could make lucrative profits by merely lending money to the Mexican government and its private sector.

While this dependent relationship was shaped and grew in the post-Camacho presidential period, the US government was also developing a dependence on Mexican resources. During the Second World War, US farmers became dependent on cheap Mexican labor imported from Mexico under the bracero agreements. After the war, American farmers refused to give up their cheap labor and demanded that the US government ensure the availability of this essential labor force. The farmers' thirst for cheap labor only got worse over time, eventually becoming a way of life for American corporations. At the same time that this relationship was developing, the US government became increasingly dependent on Mexican oil as a dependable source of energy.

CHAPTER THREE

US Dependency on Mexican Farm Labor

THE DEVELOPMENT OF A STRUCTURE

*T*HIS CHAPTER EXPLORES THE DEVELOPMENT OF US DE-
pendency on Mexican farm labor from 1942 to 2005. It is argued that in the aftermath of the Second World War, American farmers became dependent on inexpensive Mexican agricultural labor. Over the years,
this dependency intensified and Mexican labor became an essential source
of human energy in US food production. The intensification was caused by
three factors: 1) the labor shortages during the Second World War and the
Korean War; 2) the drive to maximize profits by hiring a dependable, vulnerable, and inexpensive labor force; and 3) a labor scarcity caused by Americans
unwilling to do farm labor from the 1970s to the present.

This chapter will illustrate that after World War II, the US government
resolved agribusiness labor shortages by reaching agreements with Mexico to
supply the needed labor. As a general rule, the Mexican government allowed
a migratory labor stream to enter the United States, but did little to protect the labor rights of Mexican agricultural workers. This lack of action will
be shown to be the result of economic and political structural problems in
Mexico. The outcome of this relationship in which US farmers needed labor
and many poor Mexican farmers needed employment was the development
of an asymmetrical, symbiotic, migratory labor structure. Under this structure, US farmers were assured of a constant flow of inexpensive labor, while
Mexicans received few protections from either the US or the Mexican government to maximize the value of their labor.

ECONOMIC BACKGROUND TO
US FARM LABOR DEPENDENCY

The analysis of US dependency on farm labor begins in the aftermath of the Second World War, because at that point in history Mexicans became an essential component of the agribusiness industry. Mexicans had been migrating to the United States since 1848, and migration increased in the aftermath of the Mexican Revolution. During that period many Mexicans worked in agriculture and were an important source of cheap labor. However, because American farmers relied on different ethnic groups to perform agricultural work, Mexican immigrants were only one component of the workforce. African Americans, Asian Americans, and European immigrants also worked in agriculture (Rothenberg 2000; Wells 1981). Furthermore, farmers counted on their own family members to supply approximately three-quarters of the labor. In 1910, one out of every three people in the United States lived in farm communities, and the majority of adults were engaged in agriculture (US Department of Labor 1964a: 9).

When the US government heightened its mass production of military weapons and machinery in 1940 to supply the Allied nations with weapons to fight against the Axis powers, factory employment surged and farm people began to seek industrial jobs. From 1940 to 1944, more than 1.5 million people per year left farm employment, and this pattern continued into the 1960s, with an average of 1 million people leaving farm occupations from 1945 to 1963 (ibid., 10). Once the United States entered the Second World War, in 1941, young men also left their farm communities to enlist in the military.

The gradual movement of people from farming communities created demand for hired agricultural labor. The process began when the US government entered the war and requested that farmers increase food production to feed Americans and their allies. The US government gave farmers financial subsidies to restructure their businesses and to produce massive amounts of food (Sheingate 2001: 142). Farmers, however, faced a labor shortage because the war had created alternate job opportunities for many hired hands and motivated young men to leave the fields to join the military. To resolve America's farm labor shortage, President Roosevelt sought support from Mexico and executed the "1942 Bracero Agreement," which allowed Mexican males to enter the United States to work in agriculture (EAS 278). A few thousand men were also contracted to work for railroad corporations. The men were called "braceros" (hired arms) and were issued renewable contracts for a term of six months per contract. From 1942 to 1945, American farmers employed 167,925 braceros (Grebler, Moore, and Guzman 1970: 68).

After the war, when demand for food production slowed down, American farmers lobbied Congress to retain the bracero program. Farmers alleged that they continued to experience a labor shortage because Americans did not want to return to farm labor. The US government complied and renewed the program. The war years had introduced a labor structure that farmers were not willing to dismantle, and which they would gradually become dependent on.

Mexico was content with the arrangement, since the agreement freed the government from creating jobs in the rural sector and gave it the opportunity to focus on developing urban industrial employment opportunities. Following President Cárdenas's agrarian reform executive order, succeeding presidential administrations did little for the hundreds of thousands of Mexicans who had been given land to establish farms (López 2007). During the 1940s, more than half of Mexicans were subsistence farmers, and what they produced was for consumption and not for sale (Esteva 1987: 34). Employment in the United States therefore gave many subsistent farmers the opportunity to earn wages, which in turn allowed bracero households to receive money from abroad that they used to purchase manufactured commodities. For the Mexican economy, this was a positive development, since the braceros' earnings contributed to Mexico's expanding consumer market and thereby helped to sustain the industrial sector.

US-MEXICAN RELATIONS AFTER THE SECOND WORLD WAR

In the aftermath of World War II, US-Mexican relations improved, yet with each nation motivated by different national interests, economic relations remained distant. In their classic text *The United States and Mexico: Between Partnership and Conflict*, Jorge Domínguez and Rafael Fernández de Castro (2009) argue that from 1940 until the mid-1960s, the Mexican federal government was unaggressive in its foreign relations policies, out of recognition that Mexico was a weak nation in comparison to the United States. As a vulnerable nation, Mexico had to accept asymmetrical exchange and at times tolerate illegitimate actions taken by its dominant ally, for unless Mexicans complied, the US government could rupture all diplomatic contact. As a nation adjacent to the most powerful country in the world, Mexico came to accept that its geographical location gave it economic advantages in comparison to other Latin American countries, but also presented arduous challenges. Mexico could not pursue any foreign policy that the US government might construe as a threat to its own security, or develop economic or military alliances with rivals of the United States, as this could be interpreted to be a hostile action. If Mexico misbehaved, the US government and American corporations could retaliate with an economic embargo.

Nonetheless, Mexico's alliance with and friendship toward the United States during the war earned the respect of the US government. Mexico's "good neighbor" favorable status gave it privileges not awarded to other Latin American nations. Mexico earned two major privileges: 1) a liberal immigration policy and 2) access to financial credit. US immigration policies allowed the Mexican government to alleviate its unemployment problems by permitting Mexicans to immigrate permanently to the United States or to enter as temporary workers. Employment of Mexicans in the United States greatly benefited the Mexican economy, since workers were able to send remittances to their unemployed or underemployed relatives back home. In the financial realm, the Mexican government was able to easily obtain credit, yet this practice had both negative and positive consequences. On the one hand, credit allowed presidential administrations to live beyond their means and gradually put the nation in debt. But on the other hand, US banking credit allowed Mexico to reduce its cost of food, since the government could purchase less expensive provisions on credit and pass the savings on to the people. Following the war, Mexico annually imported large quantities of food from the United States, reselling it to the public at lower costs (Andrade and Blanc 1987).

Mexico nevertheless pursued an independent economy from the Second World War until the early 1980s, realizing that distancing itself from the United States might be in its best interests. Import Substitution Industrialization (ISI) became the economic strategy to do so. Using the funds acquired from the sale of petroleum exports, Mexico began to develop its manufacturing industry, modernize its transport infrastructure, and establish state-owned industries. The government's goal was to focus on domestic production and internal consumption. Accompanying this policy, however, was the practice of setting protectionist tariffs to shield Mexican commodities from US competition. Protectionist tariffs guaranteed a domestic market and allowed targeted industries to grow and stabilize before they competed in a global market. This was particularly important in the textile and electronic manufacturing sectors. By placing high tariffs on exported products, many of which were over 100 percent of the product's value, the government gave Mexican corporations a competitive edge. Setting protectionist tariffs ensured for Mexican industrialists that although their commodities were more expensive than imported goods, sales were guaranteed because foreign imports were unavailable (Flores-Quiroga 1998; Prebisch 1971).

Indeed, Mexico's industrial goals of manufacturing shoes, textiles, stoves, refrigerators, and luxury items such as televisions and radios were important projects that ultimately reduced imports and expanded the manufacturing sector, which led to expanding industrial employment. The problem with this long-term practice, however, was that the Mexican government disproportionately used its capital, specifically from the earnings of its oil exports, to finance industrialization at the cost of neglecting the rural sector. The Mexican government subsidized companies or entered joint ventures to establish factories or to finance telecommunication companies. At other times, the government financed state-owned agencies that produced electricity, sulfur, sugar, and cement and began to heavily invest in developing Mexico's tourist industry. When investments were instituted in the rural sector, most projects aimed to improve the irrigation system and road infrastructure affecting large-scale farmers (Austin and Esteva 1987; Prebisch 1971). The logic was to give assistance to efficient farmers who could produce large amounts of inexpensive food.

For US officials, Mexico's important substitution plans and isolationist policies during the postwar years were not a concern because the US government was preoccupied with promoting democracy and containing the growth of communism abroad. Mexico was an ally, and its economic practices did not affect the American economy as long as Mexico continued to export its oil at a fair price and did not intervene in US foreign affairs. This position, however, would change in the early 1980s.

In the postwar years, the US government was far more concerned with the Soviet Union than with Latin American economic development. The Soviet Union was not only competing for the hearts and minds of the countries in war-torn Europe, but also accelerating its control of European markets (Fredrickson 2002). To thwart the Soviet Union from increasing its dominance over the reconstruction of the European markets, as well as those emerging in Asia, the US government sought to take control of the newly created global institutions designed to repair the effects of World War II.

In 1944, forty-four Allied nations met to monitor the global economy and promote international peace by establishing three international institutions: the United Nations, the International Bank for Reconstruction and Development (IBRD), and the International Monetary Fund (IMF). The United Nations was designed to create a peaceful forum for political debate and to stop wars, while the IBRD was designed to extend financial aid and loans to war-torn countries and underdeveloped nations. The IMF held parallel charters, yet its main function was to promote international trade and set the exchange value of international currency. US currency became the standard by which nations fixed the value of their currency.

Mexico's economy was not much affected by the political changes taking place in Europe, since Mexico had minimal commerce with Europe. Its main trading partners were the United States and Latin American countries, with oil being Mexico's main export commodity (Pemex 1958; Randall 1989). Although the US government was not necessarily supportive of Mexico's import substitution development plan after the war, it did not interfere. Mexico's economy was doing well, and its stability was in the best interests of the United States. The two main areas where the economies of both nations intersected were in the commerce of oil and agricultural labor.

THE BRACERO AGREEMENTS

By late 1946 the US government did not need American farmers to continue producing and exporting large quantities of food (Sheingate 2001: 119). The reconstruction of Europe was progressing well, and most nations were producing their own food. The US government was thus faced with the problem of helping American farmers scale down their production to peace-time levels. Farmers, however, were reluctant to do so. They had become accustomed to planting as much as they wanted, selling a large percentage of their food to the government, and receiving government assistance to market their crops.

To reduce the federal government's agricultural subsidy expenses, the US Congress would gradually remove the Department of Agriculture and the US Employment Service from managing most aspects of the bracero program. On November 15, 1946, the Mexican government was informed that the current bracero agreement would be terminated in four months and various changes implemented. The most radical change would begin in February 1948, when farmers were to be in charge of setting wages, recruiting workers, and paying for transportation and housing costs (EAS 1684: 1215). The US government also informed Mexico's Foreign Affairs Office that set wages were no longer to be guaranteed after March 1, 1947, and that the bracero food subsidy program would be terminated at that time (EAS 1858: 1225, 1230). During the war, the US government gave farmers food subsidy stipends to help defray the cost of feeding the braceros. Without the food stipends, the cost of meals provided in the camps would rise, since the government did not expect farmers to continue the reduced meal program. Braceros were now to be charged for the entire cost of their meals. Mexican officials protested against the unilateral changes in the program's administration. They complained that the new system was highly vulnerable to abuse, and that without the intervention of US agencies, employers would not pay braceros a fair wage or offer humane housing.

The only concessions the US government gave Mexico were related to undocumented workers and the investigation of work-related abuses in Texas. It agreed to permit some undocumented workers living in the United States to apply for bracero employment if they self-reported to immigration officials, as well as to enforce the banning of Texas counties where numerous work-related abuses were reported against farmers by agricultural workers. In the latter matter, the Mexican Consulate Office was assisted by US agents to investigate complaints and require farmers to compensate workers if they had been defrauded from their wages. The US government accepted the ban against Texas farmers, but requested that consular agents eventually recertify counties that were in compliance.

Although the Mexican government was against the new arrangement, its president, Miguel Alemán, chose not to terminate the program. His main response was to publicly castigate the US government and disclose the unfairness of the contracts (Pemex 1988: 266, 269). After this apparently shallow public spectacle of defending the workers failed to convince the US government to reconsider, President Alemán did not enact any major retaliatory policy to reverse the direction the program was taking. The president could have halted the program in 1948, when the main policy changes took effect, but this did not happen because it was in Mexico's best interests to maintain its good neighbor standing.

WHY MEXICO CONCEDED

When the US government unilaterally changed the terms of the bracero program, Mexico was in the process of negotiating a loan for US$150 million to improve its railway and transportation infrastructure. Placing conditions upon the US government to renegotiate the bracero program would not have been a strategic policy (ibid., 273). The reality was that Mexico needed an infusion of cash, and in exchange, it was willing to concede to American demands for a cheap and docile labor force. During President Alemán's term in office, from 1946 to 1952, the Pan American Highway was under construction and needed further financing to be completed. The project, which would connect the United States, Mexico, and Guatemala, had begun during World War II as part of a binational security measure. Moreover, the Isthmus Highway was also under way, and its completion was critical, as it was projected to reduce the amount of time required to transport oil from southern Mexico into the export centers. Both projects were vitally important to the Mexican economy and beneficial to the movement of people and commodities across the US border.

Alemán also needed cash to finance the extraction of oil recently discovered in the state of Tamaulipas and along the US-Mexican border, with the largest finds in the city of Reynosa (ibid., 265–266). If the president had forcibly challenged the bracero agreement, negotiations to obtain loans from the US government might have stalled or ceased altogether. Moreover, Alemán also faced a major challenge on another front that he needed to handle strategically. US officials had once again requested that the Mexican government reconsider the reentry of the oil companies into Mexico. Alemán was not prepared to open these discussions, but he had to be diplomatic to ensure that the loan agreement talks continued.

During January 1947, US ambassador Walter Hurston congratulated President Alemán for the oil discoveries and agreed to arrange for loans to be granted to Pemex (ibid., 258). This gesture, though presented in good faith, had conditions attached to it. The ambassador diplomatically introduced the idea that it might be time for Mexico to nullify its oil nationalization decree and permit Americans to invest in Pemex. Soon after, a delegation of US congressmen and businessmen met with Alemán to discuss expanding US-Mexican commerce. US president Harry S. Truman then visited Mexico to announce that 119,000 undocumented Mexican laborers would be allowed to remain in the United States and change their status to authorized bracero workers (ibid., 259). Both nations were aware that during the war years, thousands of Mexicans who were not certified to work in the United States had entered illegally and were currently employed in US farms.

Although the US diplomatic visits demonstrated goodwill on the part of US representatives, President Alemán was not prepared to initiate a campaign to change Mexico's petroleum laws. Instead, he devised a plan to quietly allow Americans to invest in the oil industry, without disclosing the terms of the agreement. In 1948 he announced that Mexico had received an infusion of US foreign aid that would be used to begin a series of oil-drilling projects throughout Mexico (ibid., 267; see Morales 1992). What he did not reveal was that Mexico had accepted US loans and made financial agreements to allow American corporations to drill oil in Mexico for a percentage of the profits. The corporate agreements were not publicized since the Mexican Constitution prohibited foreigners to explore, drill, and invest in the oil fields.

While receiving financial assistance from the United States was clearly a diplomatic and strategic victory for President Alemán, he failed to exert the same type of diplomacy when negotiating on behalf of Mexico's poorest people—the farmworker families. In exchange for fruitful financial gains and remaining in good standing with the US government, the president sold out the braceros by accepting a labor agreement that severely downgraded the value of their labor. By not terminating the agreement, the president was assured of the constant flow of their remittances into the Mexican economy. At the macrostructural level, the president's action appeared to be astute and financially logical, but at the micro level his lack of assertiveness contributed to the suffering of a vulnerable population.

THE SUFFERING OF THE BRACEROS: THE DEVELOPMENT OF A STRUCTURE WHERE NO ONE CARES

For Mexican braceros, the bracero agreement revisions caused severe hardships. When the US government turned the program over to employers in 1948, it removed US Employment Service oversight to ensure that workers received what they were owed. Under the new agreement, the contract was between employer and employee, and it was the responsibility of the bracero to obtain his pay. The worst problem, however, was that the US government no longer guaranteed a minimum wage, which during the war years was set at 30 cents an hour (EAS 1858: 1230). Wages were now to be paid based on the prevailing wage in local agricultural labor markets. The problem with this arrangement was that farmers set the prevailing wage for agriculture; therefore, if braceros were brought to an area where there was an oversupply of agricultural labor, the daily wage could be less than the daily cost of living. Under the new agreement, braceros had to pay farmers for their food and lodging; therefore, if their wages were low, they could potentially owe money to their employer at the end of the day.

Farm labor wages in the 1940s were not protected by the 1938 Federal Minimum Wage Standard Law because the US government did not want the cost of food production to rise. In 1948 the standard hourly minimum wage was 75 cents, yet the US Department of Labor reported that farmworkers' wages were much lower (US Department of Labor 2009). Farmworker wages varied, however, and they depended upon whether there was an oversupply of labor in a community, or if a worker was a year-round employee. The Department of Labor estimated that year-round farmworkers were paid higher wages. They generally were permanent legal residents or native-born. Migratory workers were paid less than year-round employees, and they moved across the country following the crops. In 1949, the Bureau of Agriculture estimated that foreign-born Mexicans and Mexican Americans composed more than half of the agricultural workforce in the United States, with their numbers estimated to range between 1 and 2 million. Most farmworkers were believed to be working legally in the United States, and only around 300,000 were estimated to be undocumented (US Department of Labor 1957a: 35).

The migratory labor force was believed to be composed primarily of undocumented workers.

The prevailing farm labor wage therefore depended upon whether the local labor force, the migratory population, and the braceros constituted an oversupply of labor. In 1949, the US Department of labor reported that the agricultural daily average wage ranged from $1.05 to $1.75, with the workday lasting longer than eight hours (ibid., 38). Based on an eight-hour day, the hourly rate ranged from 13 to 22 cents an hour, which was significantly lower than the bracero wage during the war years.

THE KOREAN WAR: A LABOR SHORTAGE

A series of events took place from October 1948 to June 1950 that led the US government to reconsider its treatment of the Mexican braceros. Mexicans placed public pressure on President Alemán to close the bracero program, Pemex reported unprecedented earnings, and the Korean War broke out, causing the US government to recruit a larger bracero labor force.

On October 17, 1948, Mexican newspapers reported an incident involving braceros in El Paso, Texas, that caused great humiliation to Mexico, leading influential citizens to demand that the bracero program be terminated (*New York Times*, October 17, 1948, in Craig 1971: 69). In Mexico, Texas farmers were known to pay the lowest wages in the United States, and to offer the worst housing conditions to Mexicans. It was common for farmers to pay braceros a few pennies for an hour of work, and in many counties farmworkers were only given tents as housing (Galarza 1964). A year earlier, in April 1947, most Texas counties were placed on a temporary ban from receiving bracero shipments (EAS 1858: 1225).

When news spread that abusive conditions continued in Texas, President Alemán asked the US government to investigate the working conditions along the border and in October 1948 prohibited Mexicans from registering in the El Paso center. On October 17, Texas farmers became very angry when they learned that they could not pick up workers in El Paso. Many enraged farmers went to El Paso and called out to Mexicans to defy the border agents, trying to persuade the Mexican workers into crossing the border. When Mexican guards refused to allow anyone to cross, a riot broke out on both sides of the border between the farmers, Mexicans who wanted to cross, and the US and Mexican border guards. The El Paso incident caused a public furor in Mexico, and a few months later Alemán warned the US government that if the treatment of the men did not improve in Texas, the entire state would once again be banned.

This incident coincided with Mexico's financial prosperity, giving President Alemán the economic confidence to renegotiate the terms of the bracero program and, if needed, temporarily suspend the program. Without the bracero remittances Mexico's economy would certainly be strained, yet this financial loss could be counterbalanced by the gains in Mexico's oil industry. In 1949, Pemex reported that its earnings had more than doubled in the previous three years and worldwide demand for Mexican oil was multiplying. For example, in 1946, Pemex's total earnings were 570 million pesos and by 1949 had grown to 1.2 billion pesos (Pemex 1958: 299). This was great news, which immediately led Alemán to announce that Mexico would use its earnings to stimulate the economy (Pemex 1988: 275). The government would double the budget of Nacional Financiera, an agency established to award grants and loans to industrial capitalists who in turn were expected to create jobs (Meyer, Sherman, and Deeds 2007: 562).

Mexico's financial prosperity occurred at the same time that the US government was preparing to enter the Korean conflict, in the summer of 1949. In preparation for the war effort, the US government signed a new bracero agreement on August 1, 1949, with additional amendments added to the agreement once the war broke out on June 12, 1950 (TIAS 2260: 1048, 1142, 1147).

The US Congress was highly dissatisfied with the agreement, but it finally accepted most of the demands on July 12, 1951, and enacted Public Law 78, changing US migratory labor laws (65 US Statutes at Large: 119). From 1949 to 1954, the annual number of braceros entering the United States ranged from 107,000 to 309,000 (US Department of Labor 1957a: 56). Direct employer hiring was terminated and the management and oversight turned over to both governments (TIAS 2260: 1051). Mexican Foreign Affairs agents were responsible for recruiting workers and negotiating contracts, while the US Employment Service was in charge of transporting and registering the men. Employers were responsible for paying round-trip transport from the men's homes to the site of employment. The men were to be housed by the US government until they were placed in their employer's camps, and the camps could not be overcrowded. Each man was to have his own cot or bed and could not be charged extra for this service. The Mexican Consul and the US Employment Service were both responsible for inspecting the camps to ensure habitable conditions. Furthermore, for the first time the braceros were allowed to elect representatives to voice their complaints and negotiate the prevailing wage, if conditions changed (ibid., 1097).

Although the US government conceded on most demands made by Mexican agents, it refused to set a minimum hourly wage. It did assure Mexico, however, that the prevailing wage could not be lower than the cost of living (i.e., lodging and food). The US Department of Labor estimated that most braceros would be paid around 50 cents an hour for an eight- to ten-hour workday (US Department of Labor 1957a: 39, 38, 60). Mexican officials were also able to compromise on the location of the registry centers, an issue that employers fought against, since they wanted all registry centers to be located at the border to reduce their transportation costs. Their logic was for the men to share the costs by paying for their transportation from home to the border registry centers. Registration centers continued to be located along the border, yet many new centers were opened in the interior of Mexico, specifically in areas of high unemployment (ibid., 57).

One area where Mexico was unable to make any inroads was in its request for the US government to impose sanctions upon employers who hired unauthorized Mexican workers and violated Public Law 78 (Craig 1971: 94; Durand 2007: 33). This became the Mexican agents' main attempt to ensure that the prevailing wage reached 50 cents an hour. Mexican agents requested that farmers be required to obtain employment permits in which they agreed only to hire authorized workers (i.e., immigrants/permanent legal residents, native born, braceros), and would be fined if they were found to be employing undocumented people. The agents' logic was to protect the braceros and prevent their shipment to farm communities where undocumented labor was available. Under Public Law 78, braceros were not to be shipped to any community where the labor force was sufficient to meet local needs, since their presence would create an oversupply of labor and bring wages down. The US Congress refused to comply with this request, but it did agree to adjust the status of undocumented workers to braceros if the men chose to register (TIAS 2260: 1050).

The US government's lax attitude toward undocumented workers was logical. By not forcing farmers to obtain employment permits, they could hire anyone they chose. If farmers did not want to hire braceros, they were free to employ less expensive undocumented labor. Small-scale farmers who were unwilling or unable to provide housing or pay for the bracero transportation fees were better off employing locals or undocumented labor, since these workers were already living in the United States. In general, only large-scale and mid-scale growers were able to pay the fees associated with the program. Each bracero cost his employer a transportation fee of fifteen dollars that was not refundable, plus additional administrative fees, including workman's compensation (domestic farmworkers were excluded from workman's compensation) and lodging when the men were in the registry centers (TIAS 2260: 1061; US Department of Labor 1957a: 60). The US Department of Labor was well aware that hiring domestic or undocumented workers could substantially reduce these costs.

The refusal by the US Congress to impose employer sanctions for hiring undocumented farmworkers was part of a larger policy to give farmers the freedom to pay the lowest wages. For farmers, the entry of large numbers of undocumented workers into their communities was good business, as these laborers did not have any protections and their presence could be used to lower the prevailing wage when local labor was sufficient but demanded higher wages. Local wages could be easily manipulated, because the US government failed to impose a minimum agricultural labor wage and allowed supply-and-demand policies to determine wages. The US government at this time did not attempt to count the numerical size of the undocumented agricultural workforce (US Department of Labor 1957a: 35).

1954: A BAD YEAR

During the Korean War, US-Mexican relations were diplomatic and fared well. The commerce of labor and oil flowed freely from Mexico to the United States, and the income from these two sources helped the Mexican economy immensely. The bracero program and the booming oil industry had greatly contributed to Mexico's prosperity. Braceros were sending around US$300,000 back to Mexico each year (Craig 1971: 102). These remittances helped the rural sector at a time when the Mexican government reported that agricultural productivity was on the rise. The US government estimated that Mexico had become a self-sufficient food-producing country, and subsistent farmers were producing sufficient food for their families (US Department of Labor 1957b: 2).

Likewise, the profits from the oil industry were being used to accelerate Mexico's industrialization. Mexico's manufacturing sector was expanding beyond Mexico City to cities such as Monterrey, Saltillo, Aguascalientes, Guadalajara, Puebla, and San Luis Potosí (Meyer, Sherman, and Deeds 2007: 565). Low industrial taxes and high profits encouraged Mexican and US corporations to invest in the industrial sector. American banks, including Chase Manhattan, Bank of America, and Hanover, were investing millions of dollars in Mexico (Pemex 1988: 286, 289, 290, 299). In 1953, however, a turn of events took place that slowed down US-Mexican commerce and led to the renegotiation of the bracero program. The problem began on June 26, 1952, when Congressman Natalio Vasquez began a legislative investigation to determine the extent of foreign investments in Mexico's oil fields (ibid., 280). He disclosed to the public that for years the federal government had been allowing American companies to invest in Pemex and share part of the profits from the oil discoveries. This, he charged, violated the Mexican Constitution.

Vasquez and his allies presented the Chamber of Deputies (House of Representatives) with undisputable proof. The newly installed president, Adolfo Ruiz Cortines, denied the charges, arguing that American oil companies were indeed drilling in Mexico, but they were paid a fee for their services. When this was proved to be false, some congressmen suggested that Mexico consider altering its Constitution and allowing the reentry of American oil companies. This caused a public scandal and led President Cortines to end the previous administration's policies. Although his administration would no longer accept new oil investments, Cortines was obliged to honor the current contracts until they expired. All foreign investment contracts were to expire by 1958 (Pemex 1988: 298).

The US government viewed this unfavorably and warned Mexico that it would not issue Pemex any loans unless the president changed his stance (ibid., 330). Making matters worse, US ambassador John Cabot informed Mexican ambassador Manuel Tello on December 30, 1953, that the US Congress would not renew the bracero agreement (TIAS 2928: 353, 356). The US government planned to continue importing agricultural workers, but it would return to the prewar arrangements, giving farmers control of the program. The official reason given was that the Korean War was over and there was no longer an emergency situation requiring the US government to oversee the movement of men to the United States. Braceros would continue to be imported, as long as farmers requested their labor.

The war's closure was an insincere justification, however, since demand for braceros continued to escalate on a yearly basis. The actual motive was that the US Congress was in the process of developing procedures to reduce the cost of importing farm labor. Ambassador Cabot also informed Tello that the US Congressional Joint Committee of Agriculture was displeased with many aspects of the program and would be making changes without seeking Mexico's approval. The committee was particularly offended by Mexico's Foreign Affairs Office and its ongoing practice of blacklisting entire counties from receiving braceros when only a few towns had violated the program's requirements. But the committee was even more displeased with the Mexican government's lack of initiative to discourage men from crossing the border illegally after they were turned away from the registry centers. Ambassador Cabot informed Tello:

> The US government is gravely concerned at the continued invasion of
> its borders by hundreds of thousands of illegal migrants. In the future
> a workable procedure must be available to control this movement,
> which will soon increase in volume with the seasonal increase in agri-
> cultural opportunities in the western and southwestern parts of the

United States. Because of this urgent need the US government must be prepared by January 15, 1954 to take appropriate action. (ibid., 356)

On January 15, 1954, President Cortines responded that unless the governments of Mexico and the United States jointly renegotiated the revisions, he would issue a temporary executive order suspending the program and prohibiting Mexicans from registering (Pemex 1988: 286). The US government ignored the warning and unilaterally continued registering braceros by allowing Mexicans to cross the border into the registry centers (Craig 1971). After a few weeks, the plan did not work as expected. The registry centers were overwhelmed, and the US government informed President Cortines that talks would resume following a congressional investigation of the program.

From February 3 to 11, 1954, the House Committee on Agriculture held a hearing to determine if there were farm labor shortages, and if so, how to remedy the situation (US Congressional Serial Set 1954a: 2). American farmers and citrus growers lobbied Congress to retain the program, regardless of the war's closure. They testified in the congressional hearing that braceros were an essential component of America's farm labor force, and that closing or reducing the program would cause a labor shortage and ultimately raise the cost of food production. In sum, they argued that farmers needed more bracero labor. Their concerns were corroborated by data on farm labor employment. On an annual basis, farmers were employing more people. In 1954 the US Department of Labor estimated that the number of hired hands had increased by at least 700,000 in the last four years. In 1954, a little over 3 million people worked in farm labor, in contrast to 2.3 million in 1950 (USDA 2008: 3; US Department of Labor 1957a: 56, 1964a: 13).

Rocio Siciliano, assistant secretary of labor, testified that farmers and citrus growers throughout the country were reporting labor shortages. However, in his analysis there was an oversupply of agricultural labor in some regions due to the size of the undocumented Mexican population. The US Department of Labor did not have an actual count, but it believed the numbers exceeded the current bracero population in the United States. Siciliano stated that the border patrol apprehended one million border crossers each year, most of whom were returned to Mexico, but he believed many made it across (US Congressional Serial Set 1954b: 4). In sum, Siciliano supported the program's continuation, but he did not issue on opinion on the number of men needed to fill the shortages.

Dr. Ernesto Galarza offered opposing testimony similar to the position taken by union representatives, such as the Teamsters Union and the AFL-CIO. Galarza, representing the National Agricultural Workers Union, testified that the domestic and undocumented labor force was sufficient. The problem faced by farmers was their reluctance to pay workers a fair wage (ibid., 176, 213). He predicted that if farmers increased wages, many Americans would be willing to work in agriculture. In Galarza's assessment, agricultural employers imported braceros to control the local prevailing wage. As long as braceros were imported, farmers did not have any reason to raise local wages.

Farmers testified at the hearing, as did representatives of the American Farm Bureau Federation, the National Council of Farmer Cooperatives, and the National Cotton Council. They challenged Galarza's opinion, collectively arguing that their main objective was to ensure that US crops were planted and harvested on time. They alleged that only braceros gave farmers this assurance, since local labor and undocumented workers were unreliable. Furthermore, they proposed that most farmers encountered labor shortages during peak harvest season.

After the hearing ended, the Joint Committee on Agriculture released its report, "Mexican Agricultural Workers," on February 12, 1954. The committee concluded that representatives of the Justice Department, the Department of Labor, and the State Department concurred that the bracero program should be reinstated. Based on the hearings, it was concluded that there was a shortage of agricultural labor in the United States. A very large amount of labor was required to ensure that seasonal specialty crops did not perish. Only braceros could offer farmers this assurance. The committee disagreed with many statements made by the labor union spokesmen, specifically with the assumption that there were a sufficient number of Americans willing to work in agriculture. The committee reported:

> In the United States there are certain areas and certain agricultural crops requiring a high degree of hand labor which need agricultural labor in great volume at certain seasons of the year, but little or no such labor at other times. . . . In these areas and for these crops it is almost impossible, even in periods of general labor surplus in the United States, to obtain a sufficient number of American agricultural workers to meet the demand at peak seasons of the year . . .
>
> The committee found unconvincing the contention of the union spokesman that the presence of unemployed industrial workers in urban centers at this time or of unemployed agricultural workers in areas far removed from those affected by this legislation at this slack season of agricultural employment, is any indication that American workers willing and able to do the agricultural labor involved in this

program will be available for that purpose next summer when the peak demand for such labor is reached. (US Congressional Serial Set 1954a: 3–4, 6)

The committee did concede that there was an oversupply of possibly one million undocumented agricultural laborers. However, the members concluded that the "wetback problem" occurred only along the border, and that the presence of these workers should not deter the government from importing braceros according to demand (ibid., 3, 6). The solution to the "wetback problem" was to increase the number of bracero jobs so that Mexicans did not enter the United States clandestinely when turned away from the registry centers. The committee advised that once Congress approved the changes for the new bracero program, a joint US-Mexico committee should be formed to address illegal immigration.

On July 15, 1954, the governments of the United States and Mexico reached a compromise (TIAS 3043: 1669). The US government would continue to unilaterally control the bracero program, but it agreed to consider instituting recommendations proposed by the newly established US-Mexico Joint Migratory Labor Commission. This meant that the conditions presented by Ambassador Cabot in December 1953 would not be rescinded. Under the new agreement, the Department of Labor was authorized to inspect camps and ensure that farmers paid their workers. However, the management of the program was turned over to employers, and a set wage was not guaranteed. The registry centers would continue to be located on both sides of the border, with three centers to remain open in the interior of Mexico in Irapuato, Guanajuato; Guadalajara, Jalisco; and Durango, Durango (TIAS 2928: 384).

The Joint Committee on Agriculture had recommended to Congress that it not concede on Mexico's wage demands, because braceros in the United States earned much higher wages than in Mexico (US Congressional Serial Set 1954a: 3). Members of the committee projected that Mexico would accept all US terms because Mexico's unemployment was rising in the rural areas. If the Mexican government chose to suspend the program, the committee predicted that due to Mexico's unemployment problems, Mexican border patrol agents would be unable to stop people from crossing the border to register for the program (ibid., 5–6).

At the closure of the talks, the new arrangement did not protect the braceros' interests. The only recommendations the US government accepted from the Mexican members of the Joint Migratory Labor Commission were that the braceros would be able to continue electing labor camp representatives to negotiate working conditions, and that workers were guaranteed a minimum daily wage of US$2 for the first forty-eight hours of work completed (TIAS

3054: 1795, 1798). This daily wage amounted to 25 cents an hour, which was half of what they received during the Korean War.

Moreover, three months after the joint commission was established, the US government began deporting undocumented workers. From June 1954 to April 1955, deportation raids were carried out in the southwestern states, with most taking place in California. Oddly, the raids did not take place along the border, where members of the Joint Agricultural Committee had reported that there was an oversupply of labor. Most raids took place in the cities, and not in the rural areas, where it was most likely that undocumented agricultural workers resided. Based on the pattern of the raids, the US Congress appeared to be concerned solely with deporting Mexicans who had left agricultural occupations for industrial jobs (see Samora 1970). The deportations ended in April 1955, when the US-Mexico Joint Migratory Labor Commission enacted a treaty to deter and apprehend undocumented migrants. Both nations concurred that the best solution was to move the bracero registry centers 160 kilometers into the interior of Mexico to discourage undocumented migration. Mexico also promised to deter undocumented migration by prohibiting Mexicans from riding on the rooftop of trains en route to the border (TIAS 3242: 1017). Poor laborers unable to afford a train fee reached the border by using this free transport method.

The number of braceros entering the US increased from an annual flow of 309,000 in 1954 to 445,197 in 1959 (Craig 1971: 130). The number of farm laborers working without a contract that had entered the United States without authorization was uncertain, since the US Department of Labor was not commissioned to maintain such records. In 1957 the department estimated that in most regions, farm labor wages had fallen or remained the same. The average wage was 50 cents an hour, which was the same wage paid to the braceros in 1950 (US Department of Labor 1957a: 80).

FARM LABOR HOUSING ALARMS AMERICANS

The lives of agricultural workers in the United States worsened in the coming years. The federal government chose not to intervene, leaving matters to local farmers (Taylor 1975). Eventually, the lack of regulation led Americans to take notice and begin demanding change. The most egregious offense was the conditions in farm labor housing, including the bracero camps. Housing conditions deteriorated toward the end of the Korean War, when the US Congress began selling federally owned farm labor housing. During the war years, the US government expanded the agricultural labor camp housing program that had begun in the 1930s. Labor camps were built

Bracero processing center in Monterrey, Mexico, c. 1956. Leonard Nadel Bracero Photographs, ID no. 2004 0138 09.16, Archives Center, National Museum of American History, Smithsonian Institution.

in communities where labor was needed but low-income housing was insufficient (US Department of Labor 1957a: 61). Government-owned buildings were modest and clean. By 1955, all federally owned labor camp housing was sold, and the government stopped inspecting these camps. By law only bracero camps were to be inspected, since the US government was required to do so based on the bracero agreement.

When the federally owned labor camps were sold, most housing was converted into rental apartments available to the public for general use. In many towns where camps were converted into apartments, rents went up, and many farmworker families could not afford to live there. The only alternative for these farmworkers was to pitch a tent on their employer's property and pay the required fee. Furthermore, most housing camps that were purchased by farmers and used for labor fell into disrepair within a few years, since federal law did not require their upkeep.

In 1956, US president Dwight D. Eisenhower established the President's Committee on Migratory Labor. The committee's mandate was to assess the housing situation of farmworkers following the sale of the federally owned labor camp housing. The committee was also to conduct a nationwide survey of the type of housing available to agricultural workers (ibid., 65). The investigation was prompted by the Farm Housing Administration, which had received inquiries from the National Council of Churches of Christ.

The council requested information on the sale of nine thousand federally owned agricultural dwellings. The council's leadership was concerned that the new labor camp owners had failed to maintain the dwellings, and that in some areas the camps had been converted into nonagricultural rental property. The Farm Housing Administration corroborated the council's concerns, since the administration's records indicated that agricultural employers were not constructing new housing or applying for loans to improve or renovate the camps.

The president's committee on migratory workers, as well as the US Department of Labor, reported that farm labor housing was inadequate and did not meet acceptable health and safety standards. The main problem identified was the failure of state legislatures to establish minimum labor camp housing standards. Most states did not have any regulations and did not require camps to be inspected. The regulation of the camps was left to local

Callused hands, job qualification for braceros, Monterrey, Mexico, processing center, c. 1956. Leonard Nadel Bracero Photographs, ID no. 2004 0138 13.21, Archive Center, National Museum of American History, Smithsonian Institution.

Braceros cross the border to Hidalgo, Texas, c. 1956. Leonard Nadel Bracero Photographs, ID no. 2004 0138 12.08, Archives Center, National Museum of American History, Smithsonian Institution.

Braceros wait to leave the Hidalgo, Texas, processing center, c. 1956. Leonard Nadel Bracero Photographs, ID no. 2004 0138 12.36, Archives Center, National Museum of American History, Smithsonian Institution.

authorities, who in turn did not set minimum standards and allowed camps to be overcrowded. Because state supervision of the camps was found to be inadequate, the committee recommended "every state that has migrant labor should have regulations governing the conditions under which these workers live" (ibid., 67). The committee also found that the bracero camps were substandard and the regulations to inspect them inadequate. Based on reports of the US Department of Labor, camp inspections were infrequent because there were insufficient examiners. It was impossible for sixty agents to visit camps across the country on a quarterly basis.

The committee therefore advised the president that agricultural housing be placed under a federal minimum standard policy, requiring it to meet the same federal standards imposed upon slums and government-owned public housing. To pressure states to improve farmworker housing, the committee also recommended that federal housing laws require newly constructed farmworker housing to have indoor plumbing. At that time only California, Florida, New York, Pennsylvania, and New Hampshire had some type of labor camp sanitation policy, but none required indoor plumbing. The president's committee concluded that most non-bracero labor camp housing consisted of barracks, tents, or one-room houses without stoves or plumbing. In spite of these substandard housing conditions, employers required agricultural workers to pay expensive fees disproportionate to what they received in housing and daily wages. The average rent a family paid their employer was $6.50 a week, or $4.80 per single person (ibid., 74). Farmworkers were also expected to pay a fee for bedding and use of a propane stove, if they did not have their own. After the report was completed, federal regulations were not imposed upon the states (Shulman 1986).

THE TERMINATION OF THE BRACERO PROGRAM

By 1963, most bracero labor was concentrated in large agribusiness farms, and this led the US Congress to reevaluate the bracero program's purpose. Throughout the country, many Americans identified the program to be the main reason why farmworkers lived in deplorable conditions and were not paid a living wage. Critics of the program also charged that only large corporations hired braceros, and during the off-season they imported braceros to keep wages low (Galarza 1964).

Indeed, by 1963 the number of farms had fallen significantly and food production was concentrated in the hands of large corporations. In 1950, 5,647,800 farms were in operation; by 1963 the number had fallen to 3,573,000

(US Department of Labor 1964a: 2).[1] The US Department of Labor, however, estimated that in spite of the drop in the number of farms, demand for hired workers was increasing because the productivity of the larger farms had improved. American farmers were producing larger quantities of crops in less acreage, and farms in states such as California, Texas, and Florida had become consolidated into large industrial agribusiness estates. Large farms thus necessitated a much larger labor force to plant and harvest crops. The Department of Labor estimated that although 52 percent of farms had relied solely on unpaid family labor in 1959, the demand for hired labor had increased on the larger farms. Six percent of farm operators employed a large labor force, while 42 percent of farms hired "some labor" (ibid., 1). The number of wage workers performing more than 25 to over 150 days of labor had increased from 3 million in 1954 to 3.6 million in 1963 (ibid., 13).

The type of hired laborers working on farms also depended on the size of the farm and the type of crops grown. Large and mid-scale growers hired locals, braceros, and migrant farmworkers, while smaller farms relied on family or local labor. Furthermore, although by 1963 braceros constituted only around 10 percent of the hired labor force (ibid., 34), the Department of Labor concluded that they served a critical function in some states. Bracero labor was generally concentrated in a few states and on specialty crops. California hired approximately 59 percent (over 65,100) of all braceros, with the majority working in the citrus and lettuce industries (US Department of Labor 1964b: 58). States that relied heavily on braceros were Texas, which employed 17,700; Florida, 14,300; Michigan, 13,500; Arizona, 8,100; and Maine, 7,600 (US Department of Labor 1964a: 33). The rest were scattered throughout the nation. In general, bracero labor was concentrated in harvesting highly perishable or specialty crops. Braceros picked 72 percent of US lettuce, 52 percent of cucumber, 49 percent of sugarcane, 42 percent of tomatoes, 29 percent of citrus and fruit, 24 percent of sugar beets, 22 percent of strawberries, and 7 percent of cotton.

With bracero labor becoming highly concentrated on large agribusiness farms, members in Congress revisited the idea of possibly terminating the program. Their motives, however, reflected vastly different political positions. Some believed that the domestic labor force was sufficient to fulfill the farmers' needs, and that therefore it was a waste of federal finances to participate in the bracero program. By the early 1960s, the US government's management role within the bracero program was costly, even though employers hired labor directly. The costs had begun to rise in 1959, after the Mexican Foreign Affairs Office demanded that worker protections be instituted, or Mexico would certify the program for only a maximum period of six months and would require renegotiations after each period. On October 23, 1959, the

US government agreed to institute ambassador Manuel Tello's requirements, which led to increasing costs. The US Department of Labor was mandated to set up a system to monitor employer-employee agreements. Employers must have proof of accidental-occupational insurance, and they must also submit evidence that a percentage of the braceros' wages was sent to the Mexican Social Security Office. Furthermore, the US Department of Labor must conduct investigations of all complaints, regardless of whether the bracero that raised the grievance had returned to Mexico. A hearing must be held, and if the farmer in question was found guilty, then he must be penalized and issued a fine. The department was also asked to inspect the camps frequently and routinely, and to be accompanied by representatives of the Mexican Consulate (TIAS 4374: 2037–2043). In December 1961 the costs of managing the program escalated when the Mexican government set three additional requirements (TIAS 4815: 1081, 5160: 2022). The first was that the US Department of Labor must set standards for the vehicles used to transport the braceros from the camps to work. Mexican agents had reported that many farmers transported braceros in unsafe and overcrowded conditions, often leading to the loss of life. It was also necessary for the US government to transport braceros from the border registry centers to the farms, to ensure that transportation was adequate.[2] Farmers had been in charge of transporting braceros, and the Mexican delegation was dissatisfied with their mode of transport. Often braceros were placed in trucks, similar to how animals were packed for shipment (see Menchaca 1995). The Mexican government's main request, however, was for the US Department of Labor to guarantee that braceros would not be shipped to communities where labor strikes were ongoing. This was particularly troublesome in California, where strike activity was on the rise and Mexican braceros faced hostile crowds. Sometimes the men were forced to retreat and missed days of work. In essence, the Mexican delegation did not want braceros to be used as strikebreakers. The US government conceded to Mexico's requests and passed regulations. The most expensive concession was assuring the Mexican government that in communities where 50 percent of the local farmworkers were on strike, braceros would not be used as strikebreakers. Instead they would be transferred to another farm, or returned to Mexico, at the expense of the US government (TIAS 5160: 2029, 2032). Therefore, the new conditions agreed upon by the US government made hiring braceros very expensive, and some congressional representatives believed it was time for federal involvement to end. They preferred that agricultural labor issues be left under state control.

Other congressmen also supported the termination of the program because they alleged it was a magnet for undocumented labor. The size of the undocumented labor force was uncertain, but it was estimated to be three times the size of the bracero labor force—most likely over one million. This estimate was based on the number of agricultural workers who were rounded up and deported between 1954 and 1955 as part of Operation Wetback. Approximately 875,000 agricultural workers were deported at that time (Grebler, Moore, and Guzman 1970: 69).

Some members of Congress opposed the bracero program on humanitarian grounds. In many regions, specifically in the southwestern states, the bracero program was alleged to lower the local prevailing wage. Once braceros arrived, local laborers had no choice but to accept the prevailing wage. In 1963 the US Department of Labor reported that in all states, the hourly farmworker wage was low in comparison to the national minimum wage of US$1.25. Farmworkers continued to be excluded from receiving the national minimum wage because it was allegedly the only way of lowering the cost of food. Farmworkers were paid from 60 cents to a dollar, and by federal and state law were disqualified from receiving time-and-a-half pay for overtime work (US Department of Labor 1963: 83, 1964b: 205).

From March to September 1963, the US Congress held a series of hearings to determine whether the bracero program should be terminated (Craig 1971: 208–209). There were doubts as to whether the program benefited all Americans. Moreover, by then labor strikes had become a nuisance to Congress, generating bad press. A year earlier, César Chávez, president of the National Farm Workers Association, had organized farmworkers throughout California and successfully launched many labor strikes, bringing the plight of the farmworker to the attention of the American public (Taylor 1975). Chávez identified the bracero program as the principal obstacle to achieving justice for the farmworkers.

At the hearings, the American Farm Bureau Federation lobbied against the program's closure, arguing that the nature of the industry made domestic labor insufficient to meet demand. Only the bracero program provided the needed labor; on many farms, crops would spoil if a large workforce was not available during harvest season. The federation also estimated that in California the cost of labor would rise by over 50 percent if the program were terminated (Menchaca 1995). Representatives for the US Department of Labor testified that the cost of farm labor would increase, but not to the extent projected. According to the secretary of labor, Willard Wirtz, US-born farmworkers, immigrants, plus the large number of Mexicans who had entered the United States illegally composed an oversupply of labor, and it was no longer necessary to import new workers (US Department of Labor 1963: 81, 1964b: 59). Wirtz recommended downsizing the labor force, as this would motivate large corporations to invest in mechanization and rely less on farm labor.

At the conclusion of the hearings, the US Congress chose to allow the bracero program to continue, but it issued administrative mandates to reduce the flow of permanent legal immigrants. This was allegedly done to alleviate farm labor unemployment, as it was assumed that legal immigrants, and not the braceros, caused the seasonal oversupply of labor. To slow down the number of legalized Mexican immigrants, the US Department of Labor and the Immigration and Naturalization Bureau were asked to prolong the application process (see Menchaca 2011). Exactly how reducing the number of permanent legal residents would reduce unemployment or lead to higher wages is unclear, as in 1964 the US government allowed 264,601 braceros to enter while it reduced the number of Mexican immigrants to 32,962 from 55,253 in 1963 (Grebler, Moore, and Guzman 1970: 68; US Department of Justice 1971: Table 14). By reducing immigration visas, the most logical outcome was that the number of undocumented people would increase.

Within the US Congress, support for continuing the program met critical opposition from labor unions, the Catholic Church, and Mexican civil rights organizations. César Chávez, the most effective speaker for the antibracero coalition, argued that braceros were being used as strikebreakers and the US government was allowing American farmers to pay servitude wages. Since the passage of the Fair Labor Standard Act in 1938, farmworkers had continued to be excluded from receiving the national minimum wage under the rationale that this was necessary to reduce the cost of food. Chávez argued that no consideration was given to how this policy affected farmworker families. It was socially unjust that the burden of keeping food prices low was placed upon the poorest people in the nation. Making matters worse for farmworkers, not only were they disqualified from receiving a minimum wage, but they were only employed seasonally. On average a farmworker's family lived on an annual income of $500, which when compared to the median US family household income of $5,600 in 1960 (US Census 1962: 1) placed domestic farmworkers in impoverished conditions. Chávez argued that to begin rectifying the problems experienced by farmworkers, it was necessary to terminate the bracero program. This would immediately cause wages to rise, as without the threat of imported workers, employers would no longer be able to manipulate the prevailing wage. US president Lyndon Johnson supported the antibracero coalition. He intervened and personally asked Congress to terminate the program. On December 31, 1964, the program was shut down (Menchaca 2011: 268). Most of the farmworkers' problems continued, however. They were not guaranteed a federal minimum wage, nor were federal laws passed to force states to improve agricultural housing.

US growers complained that the program was unfairly terminated. In 1965, six agricultural corporations launched a lawsuit attempting to reopen the program. Under *Emmanuel Braude et al., Appellant v. W. Willard Wirtz, Secretary of Labor, US, et al., Appellees* (1965), the attorneys for the agricultural corporations argued that the program was abruptly terminated with insufficient time given to farmers to find the needed labor. On that basis, the farmers demanded that they be allowed to recruit braceros. Although the corporations lost the suit and the US Congress did not reverse its position, a congressional resolution was passed to allow 36,056 braceros to enter in the next two years to work on perishable crops (Grebler, Moore, and Guzman 1970: 68).

The pretext raised by farmers that they were not cautioned the program might suddenly end was not a convincing enough argument to motivate Congress to reverse its position. Congressional debates to end the program dated back to 1954, and the issue subsequently had come before Congress on several more occasions. Many farmers heeded the government's warning, sponsoring the immigration visa applications for their former braceros in preparation for the program's termination. Reforms in the Immigration Act of 1952 had facilitated the change in status of braceros into permanent legal residents. Braceros who had entered legally could obtain a green card by filing an affidavit of support from an employer or relative and paying a fee of $25 (66 US Statutes at Large: 182, 230, 239). They could also apply for legalization of their family members by paying an additional fee of $10. To assist braceros in paying the fees, the US Department of Labor was authorized by Congress to collect the fee in installment payments (US Department of Labor 1963). Many braceros took advantage of the reforms and qualified for immigration with the sponsorship of their employer. From 1955 to 1965, 458,132 Mexicans permanently immigrated to the United States; many of the immigrants were bracero families (Grebler, Moore, and Guzman 1970: 68).

Congressional representatives did not support the farmers' request that the bracero program be reinstated. On the contrary, the majority in Congress had adopted a negative position toward Mexico and favored reducing all forms of migration. Within Congress, Mexican immigration was considered a major problem largely attributable to the bracero program. Thousands of braceros were believed not to have returned home at the end of their contracts, creating an oversupply of agricultural labor (Samora 1970). It was therefore necessary for Congress to consider immigration reform, and no longer make it easy for farmworkers to obtain legal residence in the United States.

THE NATIONALITY ACT AMENDMENTS
OF 1965 AND FARM LABOR

Less than a year after the bracero program ended, Congress amended immigration law and made it nearly impossible for future generations of working-class laborers to enter the United States legally. Under the Immigration and Nationality Act Amendments of October 3, 1965, Congress limited the number of immigrants from the Western Hemisphere to 150,000 and no longer allowed poor people to qualify for legal entry. Although Mexico was not allotted a specific quota, the hemispheric limit reduced the number of Mexicans who qualified, as Mexico had to share the quota with the rest of Latin America, the Caribbean, and Canada. For poor people, legal entry became very difficult, since section 212a of the act mandated that skilled or unskilled laborers could enter only if the Department of Labor certified a labor shortage in their respective occupations (79 US Statutes at Large: 917). In 1965 only artists, physicians, surgeons, lawyers, architects, teachers, college professors, and engineers were certified as needed labor. For a Mexican to qualify for legal entry or adjustment to permanent legal residency, family sponsorship by an immediate relative was required. However, only applicants who were sponsored by a US citizen—and thus were not counted as part of the hemispheric quota—qualified to obtain their papers with deliberate speed. Permanent legal residents were authorized to sponsor a relative, but their family members had to wait for a quota slot to become available, which often took ten years or more.

The immigration restrictions affected US farmers in the southwestern states by reducing the number of agricultural workers available. In anticipation of the problems the bracero program might produce as well as the impact the immigration reforms would have on the agricultural industry, Congress chose not to launch a deportation campaign in the rural areas. Unauthorized laborers were allowed to continue working on the farms, while in the cities the Immigration and Naturalization Service (INS) was instructed to launch a nationwide deportation campaign (Samora 1970: 85). In California, raids were conducted in Los Angeles, San Francisco, Chula Vista, El Centro, and Livermore, while in Colorado, deportations were conducted only in Denver. Raids were also carried out in the Arizona cities of Yuma, Tucson, and Phoenix. In Texas, most raids were conducted in San Antonio and Port Isabel, and along the border in the cities of El Paso, Del Rio, Laredo, and McAllen. Overall, in 1968, 150,680 Mexicans were deported, and in 1969 the number increased to 201,636 (ibid., 85, 145). Of the deportees in 1969, only twenty-nine reported agriculture as their main occupation. By 1970 the deportation campaigns in the interior of the United States ended and the INS focused on stopping Mexicans at the border.

Following the termination of the bracero program, farmworker wages gradually rose throughout the United States, but they did not increase to the level the US Department of Labor had projected. The department reported that the median wage in 1967 rose to US$1, 40 cents below the national minimum, with the highest wages paid on California farms (US Department of Labor 1988: 1, 2009: 1). Wages and working conditions, however, greatly improved in California, as the growers competed for the migratory labor force. The lowest hourly wage was $1.20, in the Coachella Valley, while the highest wage was $2.45, in Ventura County (see Menchaca 1995: 127). Many growers also began to pay time and a half for overtime work. The increased wages in California, however, were instituted not for altruistic reasons, but rather as a result of competition for the best workers, and in some areas as a result of labor shortages. This allowed Mexican farmworkers to unionize and demand better working conditions. By 1975, 80 percent of the farmworkers were unionized, and employers generally paid them the national minimum wage or above. Chávez also negotiated what many had believed was impossible: year-round domestic farmworkers employed by large agribusiness corporations were given health care insurance (Mines and Anzaldúa 1982). This golden era, however, did not last long.

MEXICO'S RESPONSE TO THE END OF THE BRACERO PROGRAM: ACCELERATE MEXICO'S INDUSTRIALIZATION

On January 10, 1963, the US government closed its largest registry centers in Hidalgo, Texas, and Nogales, Arizona. Eleven months later, on December 20, 1963, US ambassador Thomas Mann informed Mexico's minister of foreign affairs, Manuel Tello, that the bracero program would not be renewed after December 31, 1964 (TIAS 5311: 307, 5492: 1804). Mexico accepted the US position and allowed braceros to continue migrating until the registry centers closed. Mexican officials were highly aware that the program was unpopular and faced an uncertain future. The events that transpired, however, indicate that although the Mexican government knew the program's closure would shock Mexico's rural economies, it did little to assist the returning men in finding employment in their local areas. On the contrary, the Mexican government used the closure of the program to negotiate a manufacturing agreement with US corporations, which today is known as outsourcing contracts.

The Mexican government's lack of concern for the economic plight of bracero families may be explained in part by an era of relative prosperity in which the expansion of the manufacturing sector was expected to benefit all Mexicans. During the presidency of Adolfo Lopéz Mateo, from 1958 to 1964, manufacturing employment increased, Pemex sales were stable, and new oil deposit discoveries in the Gulf of Mexico and along the border indicated Mexico's economic prospects were improving. Internationally, Mexico was also classified as a country in which investments and credit were safe (Flores-Quiroga 1998; Lustig 1995). For example, from 1959 to 1963 Mexico obtained over US$124 million in loans from European and US bankers (Pemex 1988: 301, 312, 316). Mexico's economic prosperity was also shared with the Mexican people. To reduce the cost of energy and food, the government instituted major structural changes in the economy in 1960. In the energy sector the federal government purchased failing electric companies, acquiring 90 percent of the companies that produced electricity in Mexico. Subsequently, the government subsidized the cost of electricity for the poorest households and lowered the price charged to companies. Likewise, to reduce the cost of farming, the government established a government-owned fertilizer company and a national seed company (ibid., 310, 311). Fertilizer and seed prices were lowered, and a seed program was established to issue free seeds to poor farmers. These policies were expected to aid subsistence farmers, including the returning braceros.

The termination of the bracero program was tied to economic trade negotiations that started in 1962. The governments of the United States and Mexico began talks to develop manufacturing jobs in Mexico that were to employ the returning braceros as well as other Mexicans. The negotiations began with tariff discussions. Under the Trade Expansion Act of 1962, the US Congress authorized the president to negotiate a trade tariff agreement with Mexico (76 US Statutes at Large: 872–903). Mexico was asked not to impose duties on materials originating in the United States and assembled in Mexico. In return, the US government would reduce tariffs of manufactured goods entering from Mexico. By placing US factories in Mexico, US corporations were guaranteed a cheap source of labor and an immediate increase in profits, since wages were projected to be more than three times lower than those paid to American factory workers (Iglesias Prieto 2001; Tiano 2006). US factories, known as maquiladoras, were to be established in Mexico under the Border Industrialization Program. Both governments projected that the factories would serve as a magnet for workers when they returned home and looked for employment. This scenario, however, never unfolded.

Negotiations between US corporations and the Mexican government took a few years to finalize, as both sides disagreed on the location of the plants. Mexico required that they be established in zones of high unemployment, while US corporations wanted the factories to be established along the border to reduce transportation costs for raw materials and commodities. By 1963, when the bracero program was becoming increasingly unpopular in the United States, the Border Industrialization Program was regarded by Congress as the solution to repatriating braceros. It was projected that if braceros were given employment in Mexico, they would leave the United States (Cañas and Coronado 2002; Miller 1981). There would be no need to deport them. The maquiladora agreements were finally concluded on September 1, 1965, and construction began immediately (*Twin Plant News*, 2008). Although Mexico entered a new partnership to establish thousands of factory jobs, it conceded on many issues that were not in its citizens' best interests. The plants were to be built along the border, workers would not be allowed to unionize, and the American corporations would decide who was to be employed. Ultimately, US employers chose to hire Mexican women rather than the former braceros. Braceros were viewed as high-risk employees because they were men accustomed to being paid in US dollars who would command higher wages and eventually organize labor unions (Tiano 2006). In the end, the maquiladora system did not create jobs for braceros, because neither government intervened to demand that the corporations give them preferential consideration.

By 1965, twelve factories had opened along the border. They were established in locations called "twin cities," meaning where there were cities on both sides of the border. By then most of the braceros either had been deported or remained in the United States working without authorization. The plants were first established in Ciudad Juarez, across from El Paso; Nogales, across from Nogales, Arizona; Tijuana, across from San Diego; Mexicali, across from Calexico, California; Reynosa, across from McAllen, Texas; and Matamoros, across from Brownsville, Texas. In all, three thousand workers were employed at these plants (Beaumier 1990: 2). The typical employee was a young woman paid a daily wage amounting to US$1.25. It was a considerably lower wage than the braceros had received, since the wage the women earned for an entire day was equivalent to around three hours of farm labor pay.

TRANSITIONING TO UNDOCUMENTED LABOR

The immigration amendments of 1965 initially curtailed undocumented migration as working-class people in Mexico learned they were no longer eligible to apply for permanent legal residency. By the early 1970s, however, the Immigration and Naturalization Bureau estimated that undocumented migration was on the rise. One million people were believed to be undocumented Mexicans, with around 400,000 of them being related to US citizens and permanent legal residents (*Congressional Record* 1973a: 14180; Martin and Martin 1994: 16; *Silva v. Bell* 1979: 6). Many family members of Mexican Americans and Mexican immigrants who were undocumented were eligible to adjust their status to permanent legal residency, but most were unable to do so because the hemispheric quota was not large enough to meet Mexico's demand.

Oddly, while the reforms of 1965 had clearly disqualified many people from working in the United States, Congress had not passed legislation prohibiting the hiring of undocumented workers. It was illogical that undocumented migration would be discouraged while Americans were not deterred from employing undocumented people. Why Congress chose not to terminate the employment magnet that attracted poor people to the United States is uncertain. In 1973 the House of Representatives introduced House Resolution 982 to address the undocumented problem, yet once again legislators failed to consider employer sanctions as a possible solution (*Congressional Record* 1973b: 31358). Representative Peter Rodino argued during the congressional debates that the best approach to reduce undocumented migration was to establish a bracero program and set a realistic permanent immigration quota for Mexico. He proposed that Mexico needed an annual quota of at least 41,707 immigration slots to reunite families (ibid., 31365). This would not only reduce undocumented migration but also resolve the employment problems faced by farmers. Due to the closure of the bracero program, working conditions had improved in the fields, but at the same time wages had risen and there was a shortage of workers during harvest season. Closing the program had served workers well, but at the same time it raised the cost of farming. At the end of the congressional session, no resolution was passed, and immigration policy remained as it was. Due to strict immigration enforcement in the cities by INS agents, many undocumented Mexicans chose to live in farm communities, where immigration enforcement was lax and employment in agriculture was available.

By the mid-1970s, the increased numbers of newcomers led to greater employment competition and declining wages in farm communities. In 1968 the average national farmworker wage was US$1.30; by 1975 it had increased to $1.80 (Sosnick 1978: 38). However, in 1976 wages began to decline and fell to $1.60, below the national minimum of $2.30 (Martin and Martin 1994; US Department of Labor 1988: 1). Undocumented migration to farm communities also led to the devastation of farm labor unions, including the United Farm Workers Union, led by César Chávez (Martin and Martin 1994). The entry of thousands of undocumented agricultural workers who seasonally followed the crops made unionization difficult, as they could be deported anytime, making it unwise for them to pay union fees. Nonunionized workers in turn made it difficult for union organizers to gain consent and demand higher wages.[3] By 1980, Chávez lost most of the union contracts he had won (Menchaca 1995).

For some farmers the flow of undocumented labor was negatively perceived. Those who opposed hiring undocumented labor had to compete for legal workers and consequently were forced to raise wages. These farmers continued to lobby for the renewal of the bracero program. Other farmers were content with the increasing flow of undocumented labor, since many of them had been unable to import workers due to the program's expense. Farmers in general, however, benefited from the flow of undocumented labor. Undocumented people were vulnerable to employment abuses because they could be reported to immigration authorities if they complained.

The US government gave undocumented people few employment protections and created a structure that pushed farmers to become dependent on an inexpensive labor force that could be easily abused. Under the law, farmers were not prohibited from hiring undocumented workers, nor were they required to pay a national minimum wage. Essentially, by not imposing sanctions on farm employers following the immigration reforms of 1965, the US government took on a position of nonintervention and allowed farmers to pay undocumented people what the market would bear. In the short term this informal labor structure may have helped farmers to save money, but in the long-term an infrastructural process was set in motion. Farmers were becoming accustomed to employing undocumented Mexican labor. At the same time, this structure was destroying the domestic farm labor market. If wages continued to fall, there would be no incentive for US-born citizens and legal immigrants to remain in farm labor; their legal status gave them better employment options. Over the next decades, most US agricultural workers continued to be Mexican, with the majority undocumented.

In 1976 the US Congress began debates to revisit the immigration amendments of 1965. Some congressmen favored increasing the number of immigration visas given to Mexico, while others favored reducing the number of immigrants. The State Department advised Congress that Mexico needed an annual quota of forty thousand permanent legal immigrant visas to ensure that the relatives of Mexican-origin people who currently qualified for residency were admitted (*Congressional Record* 1973b: 31359; Woolley and Peters 2008). A lower quota would cause a massive backlog in immigration visas. At the end of the 1976 congressional hearings, Congress chose to pass the Western Hemispheric Act. The act made legal immigration for Mexicans more difficult, as an immigration visa limit of twenty thousand was established for all countries of the Western Hemisphere (90 US Statutes at Large [pt. 2]: 2703).

With the new quota, the State Department anticipated that the number of undocumented people would rise by another several hundred thousand.

AN AGING AGRICULTURAL LABOR FORCE, 1986

In 1982 the American Farm Bureau Federation asked Congress to resume the guest-worker program. The Farm Bureau asserted that throughout the United States, labor scarcity was driving wages up and forcing many farmers to depend on undocumented workers (Levine 2004a). In California alone, most farmers already paid workers above the minimum wage, and it was projected that current wages would rise unless new labor was imported from Mexico. Farmers throughout the country claimed that the aging braceros and their children were the only legally authorized workforce that was accessible to them. Nearly twenty years had passed since the bracero program was terminated, and with the immigration laws that were passed in the 1960s and 1970s, it was highly likely that the farmers' observations were correct.

The US Department of Labor disputed the Farm Bureau's claims and concluded that the 2.5 million farmworkers already living in the United States were sufficient to meet demand. The department also challenged the allegation that most farmworkers were undocumented. The department projected that 350,000 farmworkers at most were unauthorized (Martin and Martin 1994: 19). The reality, US agents alleged, was that the recently enacted federal minimum wage laws had forced farmers to pay all workers a minimum wage regardless of their legal status. According to the agents, farmers wanted to drive the prevailing wage down by importing braceros to create an oversupply of labor. In 1981, Congress extended the national minimum wage to farmworkers, requiring that they be paid at least US$3.35 an hour (US Department of Labor 1988: 2).

With new findings by the US Department of Labor, Congress in 1984 reassessed the Farm Bureau's complaints, which claimed that in various parts of the country the labor force was insufficient to meet demand (Cockcroft 1986). The new data coincided with reports from the Immigration and Naturalization Service warning Congress that due to Mexico's current economic crisis, undocumented migration was on the rise (Bean and Lowell 2007). Congress therefore prepared itself to address both problems at the same time, since the issues were interrelated. Farmers needed labor, and Mexicans needed employment. During the mid-1980s, Mexico experienced one of its worst economic depressions. The prosperous economy that had flourished in the 1960s and early 1970s came to a shocking halt when the price of oil fell (Grayson 1980; see chapter 4).

Mexican (Rytina and Caldera 2007: 2; Simanski 2008). By 2007, when all IRCA applications were finalized, 1.1 million farmworkers and their dependents had received amnesty under the 210A agricultural provision of the act; 900,000 were of Mexican origin (Simanski 2008). An additional 400,000 agricultural workers were legalized under different programs (Castañeda 2007: 87; see *Statistical Yearbook of the Immigration and Naturalization Service 1994*: 19). In sum, IRCA certainly served to cut the dependency of US farmers on undocumented labor. This, however, did not end the farmers' dependency on Mexican agricultural labor; it only changed the legal status of their workers.

FARM LABOR SHORTAGES POST-IRCA

Most farmers had sufficient workers for several years after the passage of IRCA, and very few guest workers were hired. From 1987 to 1994, only 7,000 guest workers per year were imported. By the mid-1990s, however, large-scale farms once again needed labor and began requesting greater numbers of Mexican guest workers. The annual number of guest workers increased to 41,000 by the late 1990s (Levine 2004a: 5). As demand for imported labor continued to increase, so did dissatisfaction with the guest-worker regulations. Many farmers complained that they did not qualify to import labor because they could not provide housing or afford the guaranteed wage. Other farmers said that the process was inconvenient, since they had to prove that there was a labor shortage before becoming eligible to apply for a shipment of workers. In general, farmers disapproved of the government regulations, and many began to hire undocumented workers. In 2000 the US Department of Agriculture estimated that out of around 2 million agricultural workers employed in US farms, more than half were undocumented (Levine 2004b: 2, 10; Rothenberg 2000: 7).[4]

In 2005 the American Farm Bureau Federation reported to Congress that agricultural labor was scarce and asked for assistance. Farmers needed Congress to make more labor available to them by reconsidering the requirements for the guest-worker program. Although farmers were importing 75,000 Mexican agricultural guest workers, many were unable to afford these workers and had to either turn to undocumented labor or see their crops perish (Castañeda 2007: 86). In support of the farm lobby, US senator Dianne Feinstein of California introduced an amendment to the AgJob Bill of 2005 (Agricultural Job Opportunity Act). To resolve the labor scarcity in the fields, Senator Feinstein proposed that the current undocumented agricultural population living in the United States be given amnesty. Approximately 1.5

million agricultural workers were estimated to be undocumented at that time (Doyle 2007: 1). Those applying for amnesty would be required to remain in farm labor employment for a minimum of three years. In addition, Feinstein proposed increasing the number of agricultural guest workers allowed annual entry and streamlining the guest-worker process so that workers could be allowed entry within seventy-two hours after a farmer's application was approved. The amendment failed to pass in the Senate. Further consideration of the amendment died in the House of Representatives, as the majority of its members did not support amnesty or liberalizing the guest-worker program. Opponents of the resolution argued that giving amnesty to workers was ineffective, because once people became permanent legal residents, they turned to industrial or service occupations. Congress therefore chose not to intervene in the farm labor situation.

Two years later, Feinstein reintroduced the AgJob bill of 2007, and this time the bill received considerable support from the Senate. In defense of the bill, Feinstein stated on January 18, 2007, "The reality is that Americans have come to rely on undocumented workers to harvest their crops for them" (*Congressional Record* 2007: 1714). The bill passed the Senate and received considerable support in the House of Representatives, but failed to pass there. A similar bill was once again reintroduced by Feinstein in 2009, but failed to gain sufficient support to become law.

During these years, Congress was divided by the legislators' different perceptions as to what constitutes a labor shortage. They all recognized that American farmers were dependent on Mexican agricultural workers, but they were indecisive as to whether farmers needed a larger labor force, or merely wanted to import workers to control the prevailing wage paid to farmworkers.

CHAPTER FOUR

Asymmetrical Codependency following Crisis Periods

*I*N THIS CHAPTER, I EXAMINE US-MEXICAN RELATIONS from 1978 to 2013 and explore how the two nations worked together for their common good during crisis periods. The chapter's aim is to explore how US-Mexican codependency developed and to analyze the role that farm labor and Mexican oil played in this relationship. I illustrate that economic problems led the Mexican government to become more dependent on US financial resources, and that US dependency on Mexican oil and farm labor intensified in turn. Mexican oil served a critical national security function in the United States, while Mexican farm labor was essential for US food production. In unfolding this account, I will present the economic background through an analysis of the Mexican oil industry. However, a more detailed discussion of the Mexican oil industry and its role in US national security politics will be explored in the next chapter. US dependency on Mexican farm labor will be the focus of this chapter.

I argue that Mexican oil became vitally important to the national security interests of the United States following the US-Iran conflict of the late 1970s, and that subsequently US-Mexican relations improved. During this troublesome period the US government needed Mexico's alliance, or at minimum for Mexico to remain a neutral observer. This event paralleled the national security crisis the US government had faced during the Second World War. It will be demonstrated, however, that Mexico's loyalty to the United States provoked a backlash from Middle Eastern nations displeased with the political pact. Following the US-Iran conflict, Mexico's economy was destabilized, largely as a result of political disagreements with Middle Eastern oil-exporting nations, which now distrusted Mexico. Although the US government periodically intervened to stabilize Mexico's economy, it is argued that

Mexico was pressured to accept economic accords that benefited Americans but at the same time had ruinous effects on the Mexican rural sector.

THE POLITICS OF OIL: DISCIPLINING MEXICO FOR INTERVENING IN US GLOBAL AFFAIRS

In 1978 massive oil deposits were discovered in the Gulf of Mexico and in the states of Chiapas and Tabasco (Grayson 1980).[1] This transformed Mexico into the fifth-largest world exporter of oil, allowing the nation to recover its global standing following the devastating effects of the oil embargo of the 1940s through 1950s (Pemex, *Statistical Yearbook 1988*: 6–8). José López Portillo, the president of Mexico, envisioned using the profits from the oil sales to convert Mexico from a middle-income developing country into a first world industrialized nation. As part of his liberal agenda, all sectors of Mexican society were expected to benefit. The president planned to reduce the nation's poverty rate, which in the late 1970s was estimated to be between 25 and 34 percent; modernize the agrarian economy; increase manufacturing jobs; and invest in the private sector (Lustig 1995: 83; World Bank 1980: 4). López Portillo's agrarian reforms were central to his economic development plan. The president projected that Mexico must become a self-sufficient food-producing nation within one decade; otherwise, two-thirds of the nation's oil profits would be used to subsidize food imports. Mexico at that time was importing a large percentage of its grains and basic food staples, reselling it at below cost because most people could not afford to pay the market value. To increase food production it was also necessary to help subsistence farmers produce for market. Sistema Alimenatrio Mexicano (SAM) was established for that purpose. SAM extended credit, technical assistance, and crop insurance, and also gave subsidies (e.g., seeds, fertilizer). SAM stores were founded to sell food at below cost.

President López Portillo also increased public spending in education and health by 28 percent (Lustig 1995: 93). To convert Mexico into a self-reliant manufacturing nation, he accelerated import substitution in this sector and raised tariffs on imported manufactured commodities to protect domestic sales against less-expensive imported goods (see chapter 1). To benefit the entire nation, López Portillo began infrastructure improvements, constructing schools, hospitals, libraries, clinics, roads, ports, and oil refineries. And to reduce the cost of living, he subsidized energy prices by selling gasoline and oil below its value. All of these programs were financed from Mexico's oil profits and through loans obtained from the International Monetary

Fund (IMF), the US Treasury, and international commercial banks. By 1980 Mexico's external debt had increased to US$50.7 billion from $9.1 billion in 1973 (Martínez Fernández 1996: 18).

Mexico's prosperity quickly ended when the Mexican government aligned itself with the United States against OPEC and was caught in an international feud over oil prices. During the Iranian Revolution, López Portillo supported the United States after Iranians accused it of imperialism and in retaliation took American diplomats hostage on November 4, 1979. After 444 days in which neither diplomacy nor US sanctions led to release of the hostages, the world prepared for war. Matters worsened when several OPEC countries sided with Iran and reduced their oil exports to the United States. In response, Mexico demonstrated its solidarity by escalating its production of oil, providing the United States with 40 percent of its foreign exports (Grayson 1980).

After the hostage standoff ended, the Iran-Iraq war followed and greatly reduced the amount of oil on the market, causing a 100 percent increase in the price of oil (Al-Chalabi 1984). This adversely affected the US economy since Americans were the largest consumers of oil in the world. To keep prices high, OPEC needed nonmembers such as Mexico to reduce the flow of oil and follow its price scale. Mexico did neither. Instead, it released more oil and increased production of light crude oil, which was in high demand in the United States. Mexico's Isthmus and Maya-mix rivaled OPEC's high-quality Arab light, which during the Iran-Iraq war became scarce. For OPEC, Mexico had become a threat, and with its massive oil reserves, Pemex held the potential to develop into an antagonistic competitor (see chapter 5).

For the United States, it was optimal for Mexico to remain an independent oil-producing nation. To strengthen Mexico's economy, the US government increased Mexican imports, encouraged investors to establish assembly factories (maquiladoras), and extended loans to the private and public sectors. The loans, however, were issued at high interest rates, ranging from 15.3 to 18.9 percent (Flores-Quiroga 1998: 252).[2] OPEC did not welcome Mexico's growing stability and independence, especially since a parallel situation was unfolding with non-OPEC oil-producing allies of the United Kingdom. In 1981 OPEC nations responded by releasing their reserves and flooding the market. The price of oil fell by 50 percent, and the United States and other industrialized nations chose to buy OPEC's cheaper oil rather than purchase their allies' oil. Mexico failed to lower its prices, and its exports sales plummeted (Díaz 1989).

López Portillo sought immediate aid from the IMF and the United States. Since World War II the IMF had helped nations obtain credit and lent money to member nations. Most members of the IMF contributed to the fund's financial reserves annually and, in turn, received the right to draw money during a crisis. In essence, the IMF issued loans and grants and extended credit to member nations. López Portillo thus expected the IMF to help Mexico during the crisis. Mexico needed the IMF to ask US banks and the US Treasury to lower their interest rates on existing loans, and for new money to be loaned. The US government was prepared to extend credit, but not to renegotiate the interest rate on the old loans. US president Ronald Reagan also requested that Mexico sell US$1 billion in oil to the United States below cost before negotiations proceeded (Boughton 2001). López Portillo considered this request unethical, as Mexico's predicament was largely caused by the alliance it had forged with the United States against OPEC. In the meantime, the Mexican economy collapsed. Mexican banks began to fail as people panicked and withdrew their funds. On a daily basis, Mexicans withdrew US$100 million (*Congressional Record* 1982a: 24019, 24164). To prevent Mexican elites from sending their money abroad, López Portillo nationalized the banks on September 1, 1982. After prolonged negotiations, the IMF convinced the US Treasury to extend Mexico credit. This was necessary to prevent the bankruptcy of US banks. The US Congress had been against issuing new loans to Mexico, but President Reagan convinced the representatives that if Mexico defaulted, 44 percent of the capital of nine of the largest US banks was at risk, as well as 35 percent of the capital of the fifteen largest US regional banks (GAO 1997: 2).

The IMF finalized the loan agreements in December 1982. Loans totalling US$12 billion were issued, with 90 percent coming from IMF and US banks (Boughton 2001: 292, 311). Only the IMF gave Mexico low interest rates, with US creditors charging from 7 to 18 percent. As part of the agreement, Mexico was required to develop a short-term economic plan to be monitored monthly by the IMF and a long-term plan reviewed quarterly. This was part of the conditionality agreement that the IMF and the US Treasury designed to ensure repayment. The economic plan contained two monumental recommendations: reduce the deficit by imposing austerity in public spending, and sell state agencies to generate income.

Under the conditionality agreement, Mexico consented to immediately cut public spending in half. Furthermore, to generate income, a series of federal financial reforms were instituted as part of the IMF–US Treasury conditionality agreement: increase tariffs; increase taxes for fuel, electricity, and food; and raise interest rates to attract investments from domestic and foreign capitalists in Mexican financial institutions. The Mexican government was also prohibited from borrowing additional funds without the approval of the IMF (IMF EBD/82/99 1982). Although López Portillo agreed to the austerity measures, he resisted the privatization of the state agencies. He held on to the belief that the state could generate wealth for the people by retaining ownership of government businesses. At this time, the federal government owned Pemex, airports, hotels, factories, warehouses, sawmills, and water, telephone, and electric utilities. The president also opposed free trade and any policy that would obstruct domestic sale of Mexican goods.

Within a year, Mexico's inflation rose to over 100 percent, the poverty rate increased to 53 percent, the peso's value fell 100 percent, and the government was required to guarantee and assume the private sector debt if firms defaulted (Stukey 1984: 563; TIAS 10961: 4610). In 1983, Mexico's external foreign debt grew to US$93 billion, of which $30.4 billion was private-sector debt (Martínez Fernández 1996: 18).

THE ECONOMIC EFFECTS OF THE MEXICAN CRISIS ON THE US AGRICULTURAL INDUSTRY

Mexico's economic crisis was expected to affect two areas of the US economy: the agricultural industry and the employment sector. The US government expected Mexican undocumented migration to increase because of Mexico's depressed economy. Agricultural workers were welcomed, since they filled an important function in the US economy, but industrial workers were not wanted because they were expected to displace Americans in factory and service occupations.

Before the crisis, the Mexican government had imported around 36 percent of its grain and 27 percent of other food products annually from US farmers (Andrade and Blanc 1987: 220; Fox 1992: 69). The food was in turn sold to Mexican consumers at below cost in government-owned stores. However, after the crisis the Mexican government could not afford to import food, and the IMF advised Mexico against purchasing it on credit since Mexico was already saddled with an annual external loan payment amounting to around US$11 billion to $12 billion. If Mexico borrowed food on credit, its amortization and servicing costs would rise. Sixty percent of Mexico's external loan payments were going solely to pay interest (Tello 2009: 243). The IMF advised Mexico to focus on domestic production rather than importing goods and accumulating further debt. Mexico's austere trade measures were harmful to the US economy, and it was necessary for the US Treasury to convince the new Mexican president, Miguel de la Madrid, to find other means to reduce federal spending (*Congressional Record* 1986b: S9645, E2581).

US agriculture was severely affected by the Mexican crisis because it lost one of its biggest markets. Toward the end of the Korean War, American farmers had begun to depend on export markets, like Mexico's, because production exceeded national consumption. New technology such as hybrid seeds, pesticides, and a doubling of the number of tractors employed in crop agriculture propelled US productivity, and by the early 1970s large-scale farmers depended on export markets to sell what was not consumed in the United States (Sheingate 2001: 17). Between 1970 and 1980, US agricultural exports increased by 150 percent, with Mexico and other Latin American countries becoming major food importers (ibid., 197). In 1985, however, this successful pattern came to a halt, and the US agricultural industry experienced its worst economic crisis since the Great Depression.

Agricultural exports had begun to slow down in the early 1980s, but they crashed in 1985, when the economies of most nations slowed down. Many countries stopped importing American food because the value of their currency fell against the US dollar. When US farmers kept the same prices rather than lowering them, many nations either purchased less food or sought cheaper markets. US farm exports fell from $43.3 billion in 1981 to $26 billion in 1986 (Harl 1990: 31). Farmers complained to Congress that they needed relief. They lobbied for the government to increase farm subsidies as well as help farmers reduce the cost of labor. In response, the US government tripled its farm subsidies in 1985 to $31.4 billion and presented other initiatives to lower the cost of farm production (Sheingate 2001: 197).

To stimulate its Latin American agricultural markets, the US government initiated the Baker Plan in 1985. Under the plan, the US Treasury and commercial banks agreed to assist Latin American countries if they agreed to open their markets to US corporations. To qualify, nations needed to be economically solvent and in good standing with their US lenders. Mexico was considered to be a key player in Latin America, as it had the capacity to become one of the largest US food importers and possessed an excellent foreign loan repayment history. To stimulate trade with Mexico, the US Treasury and US banks lowered their interest rates, converted short-term loans to long-term, and loaned Mexico new money. Mexico, however, had to follow several conditional agreements before receiving the new money: it had to resume its agricultural trade, agree to start privatizing state agencies, lower tariffs, maintain a daily US recommended oil production quota, and join the General Agreement on Tariffs and Trade (GATT) organization. At first the IMF was critical of parts of the Baker Plan, since Mexico needed loan forgiveness rather than an insignificant decrease in interest rates. Joining GATT was also questionable, since it required Mexico to end its protectionist tariff system and allow the entry of less expensive US products. Ultimately this would slow down domestic sales and harm the recovery of the private sector. The IMF also opposed the austerity demanded of Mexico because public spending already had been cut in half and educational and health budgets reduced by 30 percent (Chávez Ramírez 1996: 115; Lustig 1995: 93). However, the IMF staff changed its position in the spring of 1985 after the price of oil fell 50 percent and Mexico experienced a series of 8.1 to 7.5 earthquakes that devastated Mexico City. In September, IMF staff advised Mexico to accept US conditions, but not to lower its tariffs more than 20 percent, as this could harm Mexico's recovery (Boughton 2001: 437).

President de la Madrid complied with US requests (Guerrero Andrade 2005: 164). Furthermore, to demonstrate that Mexico was prepared to accept US conditions in return for future loans, de la Madrid lowered tariffs much further than the United States had requested. Going against IMF advice, de la Madrid lowered most tariffs by 50 percent rather than the recommended 20 percent. He also began the process of deregulating and making US investments in Mexico more attractive. He began procedures to sell state agencies to domestic and foreign corporations and eliminated a thousand permits and licenses required to run businesses. On April 1987, after negotiations were concluded, the US Treasury arranged for Mexico to receive US$11.5 billion in new loans.

Six months later, the new US loans had to be used for various purposes, and not only to resume Mexico's food imports. On "Black Monday," October 19, 1987, Mexican corporations experienced a catastrophic setback when their assets in US stocks lost 75 percent of their worth (Ramirez 1989: 12). To help the private sector, President de la Madrid guaranteed their external debts and assumed any defaulted loan. In turn, to pay for the growing federal debt, de la Madrid reduced public spending by 65 percent, increased the price of electricity 57 percent, and laid off 25,200 government employees (IMF EBS/90/10 1990; IMF EBM/93/11 1994). He also nearly terminated farm aid and food subsidy programs for the poor. These austerity measures and economic aid given to the private sector helped to stabilize the economy. The US government was satisfied with the president's response and was prepared to help Mexico further. The support given to the private sector was viewed positively, since this allowed Mexican corporations to resume international business.

OIL AND US AGRICULTURAL TRADE: MUTUAL INTERESTS

For Mexico, its relationship with the United States during this difficult economic period was fortunate yet financially precarious. US creditors extended Mexico the finances to keep the economy functioning, but Mexico had to fulfill its conditionality agreements, specifically those requiring Mexico to continue producing large quantities of oil, regardless of whether the global price of oil was low. It would have been best for Mexico to conserve its oil at this time and reduce exports until the price of oil increased. Mexico, however, was required to maintain a high oil-production level due to the 1986 Oil Exchange Contingency agreement with the IMF (Boughton 2001: 429, 437; Randall 1989: 174).

During the 1980s, the US government purchased approximately 49 to 66 percent of Mexico's total oil exports (Falk 1987: xi, 2; Shields 2005: 45). The problem for Mexico was that the price for a barrel of crude oil hit a historic low point when the Baker agreement was about to be finalized, and it remained low for many years. In 1986, the average price for Mexican crude oil was US$11.86, which was less than half the price of what it had been worth before Mexico became embroiled in the US-OPEC oil politics of the early 1980s (see chapter 5; Pemex, *Statistical Yearbook 1995*: 33; Randall 1989: 173; table 4.1).

Although the oil negotiations were not necessarily financially advantageous for Mexico, the US government did help its neighbor recover by giving amnesty to undocumented Mexicans living in the United States. On November 6, 1986, under the Immigration Reform and Control Act (IRCA), the US Congress allowed 2 million undocumented Mexicans to remain in the United States, and to continue working if they were employed (Menchaca 2011: 282). The legalization of the Mexican undocumented population was certainly a humane project that helped stabilize the Mexican economy, but it was equally important for the US agricultural industry during a period when global sales were down and farmers needed to reduce wages (see chapter 3). After IRCA's passage, agricultural wages fell or failed to increase commensurate with the cost of living. By 1990 the average farm labor national wage was US$5.23, where it remained for nearly a decade (Levine 2009: 11; NAWS 2000: 34).[3]

In sum, the US-Mexican agreements of the 1980s were beneficial for both countries. In 1988, Mexico's economic downward spiral temporarily came to a standstill, and this reprieve benefited US farmers. That year, Mexico's purchasing power improved and its foreign imports increased to US$19.7 billion, of which $11.4 billion were for US imports. Mexico's food imports also resumed, with corn, wheat, barley, cardamoms, sorghum, and beans accounting for the majority of its total imports. US corn alone amounted to $3.9 billion in sales, with corn exports reaching 5 million metric tons, nearly attaining Mexico's precrisis import level (Burnstein 2007; INEGI 1999: 36, 40, 56).

In turn, Mexico obtained the finances needed to continue running the country, and US immigration policy helped to stabilize the nation by allow-

TABLE 4.1. MEXICO'S TOTAL CRUDE OIL EXPORTS, 1977–1994

Year	Annual Volume (Barrels in Millions)	Average Price (US$/per Barrel)	Daily Exports (Barrels in Thousands)	Daily Exports to US (Barrels in Thousands)		Historical Period
				No.	Percent of Total	
1977	73.7		203.0			Cantarell estimate increases
1978	133.2		364.0			Cantarell drilling begins
1979	194.5	19.59	532.6			
1980	303.0	31.19	827.8	545.0	66	
1981	400.8	33.19	1098.0	533.3	49	
1982	544.6	28.69	1492.1	726.7	49	December, bailout finalized, US oil swap agreement
1983	561.0	26.42	1537.0	823.2	54	
1984	558.0	26.82	1524.6	750.9	49	
1985	523.5	25.33	1439.0	751.5	52	Mexico City earthquakes
1986	470.7	11.86	1289.6	652.3	51	Mexico joins GATT
1987	490.9	16.04	1345.0	639.5	48	Black Monday
1988	478.2	12.24	1306.7	684.9	52	
1989	466	15.61	1277.8	725.5	57	
1990	466	19.09	1277.1	720.5	56	US oil swap agreement
1991	500	14.58	1368.7	765.8	56	
1992	501	14.88	1367.8	798.0	58	
1993	488	13.20	1337.1	879.3	66	Persian Gulf crisis
1994	477	13.88	1307.4	960.8	73	Financial crisis unravels

Sources:
    Annual Volume, 1977–1988: Pemex, *Statistical Yearbook 1988*: 121; 1989–1995: Manzo Yépez 1996: 48.
    Average Price, 1979–1989: Pemex, *Statistical Yearbook 1990*: 36; 1984–1995: Pemex, *Statistical Yearbook 1995*: 33.
    Daily Exports, 1977–1979: Pemex, *Memoria de labores 1978*: 1536; 1979: 1737; Manzo Yépez 1996: 43; 1979–1989: Pemex, *Statistical Yearbook 1990*: 34; 1984–1995: Pemex, *Statistical Yearbook 1995*: 32.
    Daily Exports to US, 1980–1989: Pemex, *Statistical Yearbook 1990*: 38; 1984–1995: Pemex, *Statistical Yearbook 1995*: 32.

ing undocumented Mexicans to remain in the US labor force. The remittances sent by Mexicans from the United States to their unemployed relatives in Mexico were important in preventing many families from becoming destitute. At that time, Mexican unemployment was so severe that 20 percent of those eligible to work entered the informal economy by finding creative ways of feeding their families through self-employment or working in service in exchange for housing and food (INEGI 2000b: 115). In the rural areas, unemployment was much higher, and the remittances were critical to the survival of many families. The World Bank estimates that in the rural areas, out-migration spiked by 1990, and 35 out of 100 families moved to the cities or joined their relatives in the United States (World Bank 2005: 176). Throughout the 1980s, Mexico's poverty rate remained just above 60 percent (Escobar and de la Rocha 1995: 61).

US FINANCIAL AGREEMENTS DESIGNED TO IMPROVE US TRADE

After President de la Madrid stabilized Mexico's economy following the disastrous "Black Monday," the US Treasury determined that other measures must be taken to expedite Mexico's full recovery. Nicholas Brady, secretary of the US Treasury Department, proposed lowering Mexico's foreign loan interest rates to encourage Mexicans to import more US commodities. Mexico's public and private interest debt currently amounted to 60 percent of Mexico's total annual external payments. The high interest rates charged by foreign banks and the US Treasury obstructed Mexico's full recovery and dissuaded Mexican corporations from purchasing on credit. US imports were expensive, and reselling them to Mexican consumers in a highly inflated economy left little profit for Mexican corporations. If President de la Madrid was to open the economy further and allow more US commodities to enter Mexico, then it was necessary for US creditors to reduce interest rates. This could lower inflation and promote consumer spending. Secretary Brady was convinced that lowering Mexico's interest rates was necessary to expand commerce with Mexico.

In 1989 the Brady Plan was implemented to help Mexico's recovery, as well as to stabilize the economies of other Latin American countries that were also adversely affected by the US stock market crash. Mexico's new president, Raúl Salinas de Gortari, negotiated the final agreement. Under the Brady Plan, participating debtor nations were required to open their financial institutions to foreign investors, lower protective tariffs, and remove regulations that obstructed foreign investments (IMF EBS/90/10 1990). Within a year, the Brady Plan accomplished its goals. US exports to Mexico increased by 40 percent, and Mexico's external debt fell to US$95.1 billion from $109.5 billion (Flores-Quiroga 1998: 348; IMF SM/94/41 1994: 57). Mexico was now able to pay a larger percentage of its principal because only 25 percent of Mexico's external loan payments went to pay interest.

The US government found that practicing good-neighbor policies toward Mexico benefited both nations. President Salinas in return began to privatize state agencies on a large scale, in many cases selling them below their worth. Domestic and foreign creditors were given the option of exchanging their government loans for state property. Of 1,155 federally owned agencies during President López Portillo's term, only 765 were left by 1989 (Guerrero Andrade 2005: 159). Salinas also chose to reform the treasury bond market as a means of generating income. He opened it to foreign investors and guaranteed a 20 percent interest earning on long-term mature bonds (Chávez Ramírez 1996: 70, 244). This was an incredible return. Many Mexican economists criticized the president and predicted that although the bond market would certainly attract investors and increase the flow of money, in the near future this policy was doomed to self-destruct. How would Mexico be able to pay the interest in the future if investors chose to cash in their bonds? The president did not heed the economists' warning, and the sale of high-interest bonds continued to increase over the years. By 1994, US investors owned over 70 percent of Mexican bonds (GAO 1996).

THE NORTH AMERICAN FREE TRADE AGREEMENT: REFORMING MEXICAN OIL AND AGRICULTURAL POLICIES

President Salinas, a Harvard-educated economist and member of one of the wealthiest families in Mexico, went further than former President de la Madrid in liberalizing the market and developing closer economic ties with the United States. Salinas rejected protectionist trade policies and advocated the neoliberal principle of free trade. He favored replacing GATT with a new proposal promoting free trade and eliminating tariffs within fifteen years. The North American Free Trade Agreement (NAFTA) was designed for that purpose. In 1990 talks began between Mexico, the United States, and Canada.

NAFTA negotiations initially stalled after the Mexican Congress refused to accept US conditions to privatize Pemex. US president George H. W. Bush proposed that before he agreed to NAFTA, privatizing Mexico's oil industry was necessary (Boué 2006). US attorneys informed President Salinas that Article 27 of the Mexican Constitution did not apply to future oil discoveries; therefore, selling newly constructed oil fields after NAFTA became law was not a violation of the Constitution. The attorneys also offered the opinion that post-NAFTA US investments in the oil fields were legal. The Mexican Congress rejected the proposal outright.

When the privatization proposal failed, US negotiators introduced for discussion Article 650 of chapter 4, pertaining to "Energy and Petrochemicals." They informed Mexico that President Bush was willing to support NAFTA if Mexico agreed to make its external loan payments with oil shipments rather than cash payments. This agreement was contingent upon Mexico accepting a US-imposed annual export quota, with the further condition that if Mexico could not meet the quota, it must decrease its domestic consumption of crude oil and reduce its exports to other countries (Manzo Yépez 1996: 35). Once again, the Mexican Congress overwhelmingly rejected Article 650 and pressured President Salinas to retreat from the negotiations if the US government persisted on an oil agreement. Many congressmen were insulted that the US government had not placed similar conditions on Canada, which was also a major oil-exporting nation. Salinas was in support of partial privatization, but when he met unified congressional opposition, he abandoned his personal views.

The NAFTA accord was finally executed on January 1, 1994, after US president William Clinton took office. Clinton was satisfied with Mexico's trade stipulations and accepted the Mexican Constitution's prohibition against privatizing the oil industry. Salinas had signed the accord one year earlier, conceding on many reforms affecting Mexican agriculture. He agreed to eliminate all tariffs in the agricultural sector except those pertaining to corn, beans, and dairy (IMF SM/94/41 1994). All Mexican tariffs would be eliminated within fifteen years, including the agricultural tariffs. Salinas also agreed to end farming subsidies and issue only limited assistance to farmers who produced for market, projecting that this reform would stimulate agricultural productivity and immediately force inefficient farmers to sell their property to productive farmers.

President Salinas also agreed to other free market reforms, including reducing regulations and accelerating the privatization of state agencies. Both reforms were expected to attract foreign investments. Salinas endorsed the neoliberal philosophy that the private sector could run state firms more efficiently than the government. Salinas's free market approach appeared at first to be a monumental success, as direct foreign investment increased to US$93 billion by 1993 (GAO 1996: 33).[4] In late 1994, however, IMF staff warned Salinas and the US Treasury that unless the Mexican economy received a massive infusion of cash, its collapse was imminent and would be damaging to the US economy (ibid.). A massive surge in undocumented migration to the United States would soon follow.

So what happened to destabilize Mexico during the Salinas administration? Unlike in past economic crises, the fall of global oil prices was not the cause. Mexico's oil industry was stable and remained the nation's main source of income.

HOW THE 1994 CRISIS DEVELOPED

The collapse of the Mexican economy was triggered by President Salinas's immense mistakes within the banking and financial industries. As part of his neoliberal agenda, he privatized Mexico's banking industry and recklessly increased the interest earnings for Mexican treasury bonds. He planned to generate federal revenue by privatizing the nationalized banks and to attract foreign revenue by selling treasury bonds. The IMF had advised Salinas to sell the banks in order to reduce the government's financial risks (Chávez Ramírez 1996).

In 1990 the banks were privatized and sold to Mexican citizens (Barnes 1992; IMF SM/94/270 1994).[5] Under the López Portillo administration, sixty-six banks had been nationalized and consolidated into eighteen banks, of which six were nationwide. They were nationalized to stabilize the Mexican economy when the OPEC-Mexico crisis began in the early 1980s. When the banks were sold, many Mexican economists criticized the plan because it had taken the government many years to make the banks solvent. Critics also proposed that the banks were intentionally undervalued to attract investors, and that the government would therefore lose substantial assets when they were sold. More significantly, they argued it did not make any financial sense that the government was willing to lend buyers part of the assets to purchase the banks. Many wondered, if the aim of the privatization initiative was to generate income, why then was the government going to sell the banks on credit? (Martínez Fernández 1996).

The banks were sold for US$13.5 billion, and within a couple of years 96 percent of the funds were used to pay for Mexico's external debts (Chávez Ramírez 1996: 103; IMF EBS/90/10 1990). To help the banks generate income, President Salinas began a treasury securities partnership business venture. Banks were to sell high-interest treasury bonds and notes guaranteed by the federal government, and upon maturity most investments were to be paid back in US dollars. Interest rates would range from 14 to 37 percent (IMF SM/94/41 1994: 46; INEGI 1996: 163). Within four years, the sale of treasury bonds and notes generated around US$43.4 billion for the banks; 70 percent of the bonds and notes were purchased by foreign investors, with the majority being US citizens (GAO 1996: 64, 6, 11; Girón González 2001: 278). The president had envisioned that bankers would use the treasury bond and note sales to stimulate the Mexican economy by offering credit to Mexican citizens and investing in profitable business projects.

Although the IMF supported the banking reforms, its staff criticized Salinas's securities policies (GAO 1996; IMF SM/94/41 1994). The bond and note sales were excessive, and to make matters worse, the president began a similar partnership with the states and with other types of financial institutions. The IMF warned Salinas that if a large percentage of the investors chose to cash in their bonds and notes upon maturity, the Banco de México (national bank) would not have enough assets to meet the payments. The IMF also criticized the president for forgiving the loans of several bankers who had borrowed federal money to purchase the banks. This benevolent practice was part of Salinas's modus operandi, which gave federal assistance to large-scale corporations while neglecting small- to mid-scale businesses. During Salinas's administration, 65 percent of federal aid to the private sector (i.e., credit, grants) was given to 343 corporations, while 28,975 small-scale businesses received only 6 percent (INEGI 2011: 433).

A contributing factor in the destabilization of the Mexican economy was the sale of state agencies. By 1995, Mexico had sold most of its state agencies and had few resources to generate income. The sulfur, salt, petrochemical, textile, mining, food warehouse, food processing, railway, telegram, tourism, finance, television, telephone, and taxi industries were sold (Chávez Ramírez 1996; IMF CR/9/7 2009). Only 215 agencies were owned by the state, and of these, 106 were jointly owned by foreign or Mexican entrepreneurs. Profits from these privatized agencies now belonged to investors. In 1994, *Forbes* magazine announced that Mexico had produced many new billionaires in the previous year. Mexico now ranked fourth in the world in billionaires, with twenty-four people achieving that stature (*Forbes* 1999).

Once state agencies were sold, President Salinas began to lobby the Mexican Congress to remove regulations that allegedly impeded corporations from maximizing their profits. Salinas supported the deregulation of the tobacco and tortilla industries and eliminated most license and permit fees for growing fruits, vegetables, and cacao.

Deregulation and privatization at first stimulated the economy, particularly the banking reforms. The enactment of NAFTA also appeared to boost the economy. For example, large-scale Mexican agricultural producers did well and greatly increased their sales of tequila, beer, produce, coffee, and cattle. The number of jobs also increased in the maquiladora factories, as US corporations began to open new companies in anticipation of NAFTA's enactment. These corporations were highly motivated by President Salinas's environmental deregulation initiatives in the petrochemical and paint industries. From 1990 to 1994, the number of maquiladoras increased from 1,703 to 2,085 (Mattar 1998; Wilson 2010).

However, in March 1994, four months after NAFTA took effect, the banking industry nearly collapsed after thousands of farmers and mid-scale industrialists defaulted on their loans. Mexican businesses rapidly began to fail after less-expensive US and Canadian commodities flooded the markets and made competition impossible. As the economy unraveled, foreign corporations stopped investing in Mexico and by December started pulling out their money (GAO 1996: 14, 54, 74). Mexico quickly became politically unstable, as Mexicans in fear of losing their life savings tried to withdraw their funds, only to find that banks were bankrupt. In the state of Chiapas, where the agricultural deregulation policies had devastated the economy of subsistence farmers, the *Ejército Zapatista de Liberación Nacional* declared war against the Salinas administration and began armed revolt. In Mexico City, Luis Donaldo Colossio, a populist leftist leader, declared his nomination for the presidency, promising to reverse Salinas's neoliberal policies (López Obrador 2010).

In December 1995, the banks collapsed. The main problems were the banks' inability to pay the bonds and notes, and their failure to recover the loaned credit. The banks had a liquidity deficit of over US$37 billion and also shared the treasury bond debt of $30 billion (GAO 1996: 6, 11; Girón González 2001: 278; IMF SM 94/41 1994: 59; Ros and Bouillon 2001: 727). Exacerbating matters, the federal government and the private sector did not have the funds to meet their external debt payments. In total, Mexico's foreign debt amounted to US$140.1 billion, of which $28 billion was due in principal and interest (Martínez Fernández 1996: 18). With the economy in shambles, thousands of corporations, restaurants, and retail stores went bankrupt and people lost their jobs. Salinas's free market society had failed. In fear for his life, Salinas began a self-imposed exile in Ireland.

THE BAILOUT AND THE OIL AGREEMENT OF 1995

Ernesto Zedillo, Salinas's secretary of education, succeeded to the presidency on December 1, 1994. The front-runner for the presidency, Luis Donaldo Colossio, was assassinated. Zedillo was confronted with the onerous job of restoring the nation's confidence in government, meeting the external debt payments, and bringing closure to the Zapatista Movement (GAO

1996). To do so, he turned to the IMF and President Clinton. Zedillo, like Salinas, was an economist educated in US schools. He received his PhD from Yale University.

Zedillo finalized a loan agreement with the US government in March 1995 after he agreed to use the petroleum industry as collateral. With most of Mexico's state-owned agencies sold, oil was the main asset the Mexican government had to qualify for credit. Under a US Treasury and IMF–designed economic recovery plan, the Mexican federal government agreed to reduce 40 percent of its domestic spending and implement a series of new deregulation policies that included allowing banks to be fully owned by foreign investors and privatizing the majority of government-owned industries (e.g., electricity, highways). Most of the recommendations were instituted except for the privatization of Pemex, which the US Treasury and the IMF had requested.

Although Mexico needed an immediate infusion of cash, President Zedillo did not act to privatize Mexico's oil industry, since he did not expect the Mexican Congress to accept such a radical revision of the nation's economy. It was also strategically impossible to launch a successful privatization campaign during the bailout negotiations. As part of the conditionality agreement negotiations, however, Zedillo was able to convince the majority in the Mexican Congress to accept the US Treasury's plan for the oil industry (Manzo Yépez 1996). The congressional representatives agreed to allow the US government to set the price and amount of oil to be exported for the next ten years. If Mexico refused to follow the export plan and its volume of sales declined by 15 to 25 percent, the US Treasury was authorized to require immediate payment of US loans (GAO 1996: 2). Pemex was also prohibited from accepting new crude oil export contracts and was instructed not to renew expired contracts. US consumers were to be Mexico's primary importers. In 1995, Mexico sold 80 percent of its foreign exports to the United States, a significant increase from previous years (Manzo Yépez 1996: 47; Shields 2005: 45).

In return, the federal government and banks of the United States issued Mexico a series of loans amounting to US$20 billion in direct funds. The US Treasury also became the guarantor for $28 billion in loans to be lent by Canada, the IMF, and the Bank for International Settlements. Although Zedillo refused to initiate proceedings to privatize Pemex, President Clinton accepted the agreement since failure to do so would affect the American economy. The US Treasury projected that trade with Mexico would stop and 700,000 US jobs would be lost. The United States supplied 69 percent of Mexico's total foreign imports, which amounted to 10 percent of US exports (GAO 1996: 3). Moreover, since a large percentage of Mexico's treasury bonds and notes were owed to US banks, the bailout was necessary to ensure repayment. The US Treasury had warned Clinton that US banks would be severely affected if the bonds and notes were not paid. As part of his negotiations, President Zedillo agreed to compensate American investors if a financial package was given to Mexico (ibid., 123). Basically, what was being loaned would be returned to the US economy, with interest.

Without a doubt, the stabilization plan helped Mexico's economy recover, and within one year the IMF reported that foreign investments in Mexico were safe. As part of the bailout agreement, the private sector debt and the banks' share of the treasury bond and note debt were assumed by the Mexican government. This contributed to the stabilization of the private sector and allowed capitalists to carry on business as usual (ibid., 22; Girón González 2001: 245, 249, 271). The problem for most people, however, was that the burden to pay for Mexico's external debt, including private sector debts, was imposed upon them. The common person was placed under strict government austerity policies, because reduced public spending was needed to generate federal income. This was necessary, as substantial assets were needed to meet Mexico's foreign debt quarterly payments and generate new monies to assist the private sector. During Zedillo's administration, annual inflation increases fluctuated from 135 to 254 percent (INEGI 2000b: 49). The middle class was also severely affected. It became difficult to qualify for credit, and the interest rate for credit cards inflated to at least 38.5 percent (GAO 1996: 139). The poverty level increased to 69 percent and remained in the low 60s throughout Zedillo's administration (Consejo Nacional de Población 2009: 25). Moreover, farmers displaced by NAFTA's agricultural policies were not absorbed into manufacturing occupations. The World Bank estimated that 50 percent of agricultural workers entered informal occupations, such as self-employment or unpaid service jobs, in exchange for housing and food, while others migrated to the United States (IMF CR/4/418 2004: 15).

CONDITIONS LEADING MEXICAN FARMERS TO BECOME US FARMWORKERS: AN EMPLOYMENT DEPENDENCY STRUCTURE

After the Mexican economy collapsed, the rural sector was devastated when Salinas's plans to restructure the agricultural industry failed. A few years earlier, Salinas had set a plan in motion to remove inefficient farmers from the agricultural industry; he projected that after being displaced from farming, they would be absorbed within the manufacturing sector. This plan became impossible to implement after the economy collapsed.

Salinas had envisioned that it was best for Mexico to import less-expensive food from the United States rather than continue to give state aid to inefficient farmers. If unproductive farmers were pressured to abandon farming, the cost of food production would decrease because the Mexican government would allow less-expensive food to enter the Mexican market. In essence, imported food from the United States was projected to lower the cost of food. This scenario did not unfold. On the contrary, in the aftermath of NAFTA, displaced farmers found themselves unemployed and in worse poverty than before. Seeking a better life, many farmers joined the migratory circuit to the United States or to Mexican cities.

Those who chose to migrate to the United States, however, did not qualify for legal entry, and the majority crossed the border illegally. For the US government, the large-scale migratory movement of Mexicans to the United States generated both positive and negative outcomes for the US economy. On the one hand, the entry of a large labor force created an oversupply of agricultural labor, thus contributing to lower agricultural wages, which was ultimately profitable for employers. On the other hand, the US government had to deal with a radical increase in undocumented immigrants. Mexicans from the rural sector were not the only ones who migrated to the United States. The US Government Accountability Office estimated that by 1997, the undocumented population had increased significantly and over five million workers plus their dependent families were living in the United States (Stana 1999: 6, 17).

In the United States, Mexicans who entered farm labor came to play an important function in the economy and were a welcomed labor commodity. Mexicans who preferred industrial occupations, however, were seen as taking jobs away from American citizens and in general were viewed as an unwanted and annoying hindrance to the US economy. For the US agricultural industry, Mexico's agricultural restructuring after NAFTA overwhelmingly benefited business, not only providing labor but also opening Mexico's food markets. The restructuring of Mexico's agricultural industry began when President Salinas prepared to institute reforms that would convince the US Congress to support NAFTA. Before the US government agreed to finalize NAFTA negotiations, Salinas had to create free market conditions in Mexico's agricultural industry. Salinas began to implement policies in 1990 to accommodate US requests. The US government required that Mexico purchase American farm products and not block the entry of less-expensive American food. For Salinas it was therefore necessary to reduce the number of Mexicans engaged in farming, particularly in the production of corn, since most farmers in Mexico grew this staple.

From 1990 to 1992, President Salinas passed a series of laws to remove noncompetitive small-scale farmers from the agricultural industry. Working with the US Treasury and the US Department of Agriculture, Salinas determined that the cost of Mexico's food production could be lowered if Mexico purchased most of its corn from the United States, since US corn production was less expensive than Mexico's.[6] IMF economists agreed and projected that the price of corn would fall by 38 percent and consumption would rise by 18 percent (Larsen 1993: 4). Salinas proposed to restructure the corn industry by removing the Mexican government from the corn business and allowing inefficient farmers to go bankrupt. He no longer wanted the federal government to purchase most corn harvests above global market value and then resell the corn at below cost. If Mexico could purchase less-expensive corn, then subsidizing farmers would be unnecessary.

The main problem for Salinas, therefore, was to devise a plan to remove a large percentage of the corn farmers from the industry. This was a major task, since most farmers grew corn (Guerrero Andrade 2005). Salinas and NAFTA representatives knew that removing Mexican farmers from the industry would shock rural economies, but they firmly believed this was necessary to improve food production (IMF SM/94/41 1994). Salinas's restructuring plan consisted of three main goals: to end farm aid to inefficient farmers, terminate protective tariffs, and reward productive farmers. Those who could survive these policies would do so because they were efficient farmers.

President Salinas began the reforms by restructuring Mexico's communal land system, which is known as the *ejido* system. To make subsistent farmers less dependent on the state, the president obtained the majority consent of Congress in 1992 to terminate the *ejido* system and end farm aid to unproductive farmers (Harvey 1998: 187). Article 27 of the Mexican Constitution was reformed, allowing owners of communal lands to privatize their property. Since 1917 the federal government had prohibited communal lands from being sold, to ensure that *ejidatarios* had land to grow crops. Salinas's structural plan allegedly was designed to convert the *ejidatarios* into private-property owners who would obtain credit by using their property as collateral. Though this was the rationale, Salinas had undermined these farmers' ability to obtain credit. Two years earlier he had reformed the banking system and partially privatized Bancorural, the government's agricultural development aid bank. Under new private management, credit was limited to farmers with substantial assets (Carton de Grammont 2000; Guerrero Andrade 2005). To determine who was eligible to obtain credit, farmers were classified under three categories: 1) producers with capacity who can compete in the market, 2) producers with potential who are identified to have the means to become productive, and 3) those without any potential and at high risk to default on government aid (Carton de Grammont 2000: 73). *Ejida-*

tarios and small-scale farmers were classified under category 3 and disqualified from obtaining government credit. Those under category 2, "producers with potential," constituted 35 percent of Mexican farmers; they might be eligible for government credit. Only farmers with capacity, category 1, were deemed a safe risk to receive government loans. This sector constituted 15 percent of Mexican farmers, most of whom were large-scale farmers.

To further destabilize the *ejidatarios*, President Salinas ended most farm subsidies in 1993. He announced that all subsidies would end in fifteen years, but in the meantime, farmers would be given financial assistance based on the amount of acreage they had under cultivation.[7] Subsidies would be awarded to help pay for inflation costs. In 1993, only large-scale farmers received subsidies, as well as those growing specialty crops for export, such as cotton, tobacco, sugarcane, coffee, and sorghum (IMF SM/94/41 1994). Nonetheless, even successful farmers complained to the government that the assistance they were given was insufficient. Farm production costs had increased by 18.6 percent, yet farmers received only a 12 percent inflation subsidy (Guerrero Andrade 2005: 181, 194). By 1999, most farmers did not receive any aid, and across the country only 378,000 hectares were covered by a subsidy.

When NAFTA took effect, in 1994, President Salinas's tariff policies pushed most noncompetitive farmers out of agriculture. Nearly all agricultural tariffs were removed that year, and less-expensive US and Canadian crops flooded the market (Guerrero Andrade 2005: 166). This was a major setback for farmers, since they received little to no financial assistance from the government to compete with US agribusiness firms. At this time US farmers received $7.9 billion in direct government payments, and by 1999, payments had increased to $21.5 billion (USDA 2015a). Salinas's tax and privatization policies also made farming more expensive and increased the difficulty in earning a profit. For example, when the government began to lease public waters (streams, rivers) in 1992, many farmers could not afford to pay a fee and were doomed to fail. Likewise, in that same year the government sold Fertimex, Mexico's main fertilizer company, and the price of fertilizer rose (Chávez Ramírez 1996; López 2007).

These agricultural policies were allegedly designed to foster competition in the agricultural industry and reduce the price of food by eliminating nonproductive farmers. Indeed, the number of adults reporting agriculture as their main occupation fell considerably after NAFTA was enacted into law. In 1970, 40.6 percent of adults age fifteen and above reported agriculture as their main occupation, but by 1997 the percentage had dropped to 20.8 percent (INEGI 2000b: 114).[8] Whether reducing the number of farmers was a positive outcome for the Mexican economy is debatable, since the price of corn did not fall as projected.

From 1995 to 2000, the price of corn rose, leading to radical price hikes in Mexico's main food commodity—the tortilla. According to Mexican congressional findings, the price of tortillas increased by 279 percent, which caused Mexico's inflation to rise (Cruz Miramontes, Cruz Barney, and Aguilar Mendez 2009: 319; see USDA 2008: 10). Economists disagree on why the cost of corn failed to fall as projected by the IMF and the Mexican government. Some economists propose that this happened because the Mexican-owned firm GRUMA S.A. gradually monopolized the corn industry in Mexico, and without government regulation was allowed to repeatedly increase prices (Cruz Miramontes, Cruz Barney, and Aguilar Mendez 2009: 486).

Other economists believe that this was a result of a few US farmers monopolizing food exports to Mexico (Guerrero Andrade 2005; López 2007; USDA 2004). US corn exports increased 240 percent from 1995 to 2004, accounting for approximately 25 percent of Mexican corn consumption (USDA 2004: 3). One US corporation, Cargill, owned 40 percent of imported corn sold to Mexican consumers and 10 percent of corn produced on Mexican soil (Guerrero Andrade 2005: 195).

AFTER NAFTA: WHY THE MANUFACTURING INDUSTRY FAILS TO ABSORB DISPLACED MEXICAN FARMERS

After NAFTA, Mexico's manufacturing plans did not develop as projected. The private sector failed to produce sufficient manufacturing jobs to absorb the displaced agricultural workers. In the 1990s, manufacturing employment remained at the same level as twenty years earlier, ranging from 24.5 to 25 percent of the total workforce (INEGI 2000b: 115). For decades, most job growth was in service occupations, which generated less income than subsistence farming. The displacement of farmers and hired farm laborers was massive. The US Department of Agriculture proposes that from 1991 to 2000, the number of Mexicans engaged in agricultural production fell from 4.3 million to 3.4 million (USDA 2004: 6). Likewise, the World Bank estimates that agricultural employment shrunk by half from 1995 to 2003 (World Bank 2005: 2:119). Mexican government estimates are even more startling. The Instituto Nacional de Estadísticas y Geografía (INEGI, National Institute of Statistics and Geography) found that in 1984, 8 million Mexicans reported agricultural work as their main occupation; by 2006 the number had dropped to 6 million, and by 2009 to 5 million (INEGI 1984: 9, 2009a: 1310, 2009b: 26).

Mexico's inability to absorb displaced Mexican farmers in manufacturing cannot be attributed to the reluctance of US investors to open factories in Mexico. On the contrary, US firms have historically owned 90 percent of Mexican maquilas and produced from 50 to 80 percent of Mexican manufactured exports (Cañas and Gilmer 2007; Moreno-Brid and Ros 2009: 184). The problem is that maquila production has become very efficient and fewer workers are needed to produce the same amount of goods.[9] Mexican industrial workers are very productive; they work fast and endure long hours. Thus from 1995 to 2002, Mexican manufactured exports increased by 80 percent, yet this did not translate into job creation. The IMF found that maquila plants have become increasingly efficient over the years and have higher productivity per worker (Kose, Meredith, and Towe 2004: 15). This efficiency unfortunately does not lead to significant increases in the number of manufactured jobs created on an annual basis. In 1995, 2,939 plants were in operation, and by 1998 the number had increased to 4,234 (Domínguez and Fernández de Castro 2009: 139). In terms of the number of employees, 644,000 Mexicans worked in the maquilas in 1995 and by 1998 the number had increased to 1 million. This growth, however, was not sustained and has fluctuated over the years. After 9/11, when the US economy was shocked by terrorist attacks on US soil, manufacturing employment in Mexico actually fell. In 2003, 800 maquiladoras were closed and 300,000 people lost their jobs (Fabens 2013: 4). This altered maquila employment, and manufacturing jobs fell to 19 percent of the total workforce from 24.5 percent in 1995 (World Bank 2005: vol. 5, 115). In 2008, manufacturing recovered slightly to 23.2 percent of the Mexican workforce, but it continued to be lower than the 1970s levels (INEGI 2009c: 14).[10]

The point is that Mexico's manufacturing sector has not grown sufficiently to employ the growing number of Mexicans displaced from agricultural occupations, and many workers chose to migrate to the United States in search of employment. Their absorption within the United States was possible since US agriculture continued to prosper.[11]

US CONGRESS MANAGES MEXICAN FARM LABOR MIGRATION: CLOSING AND OPENING THE GATES

Throughout the 1990s, the US Congress revisited immigration law, but it did not pass amnesty legislation targeting all Mexican people. When adjustment policies were instituted, the legislation only applied to farmworkers and their families. That was a sensible consideration, since American citizens by choice preferred not to work in farm labor. Allowing a trickle of farm laborers to adjust their status on an annual basis served the interests of the agribusiness industry and the welfare of the nation as a whole (see Levine

2004a). Therefore, although the majority consensus within Congress was to stop people from fleeing Mexico's disastrous economy, provisions for agricultural workers were needed because farm labor continued to play an important function in the US economy.

When the Salinas economic crisis of 1995 occurred, the US Congress had already set in place the political infrastructure to ensure that US growers had sufficient labor. The 210A provision of IRCA in 1986 and the Immigration and Nationality Technical Corrections Act of 1994 allowed growers to replenish their labor force. The 210A program allowed undocumented agricultural workers to apply for legalization (SAW applicants) and also set the legal infrastructure to allow farmworkers who arrived after the passage of IRCA to adjust their status (sec. 210A in 100 US Statutes at Large [pt. 4]: 3417). These policies were designed to give employers the opportunity to sponsor workers whom they believed were valuable employees. Under the program, farmworkers and their families were allowed to adjust their status. The initial agreement under IRCA was set at a maximum of 350,000; however, a provision was appended allowing the Department of Labor and the attorney general to adjust the number of people who were legalized under this program. The 210A program was to be terminated by 1993, but could be continued under the discretion of the attorney general. Mexican families not engaged in farm labor did not qualify for adjustment under the 210A program.

On October 25, 1994, when it was apparently clear that Mexico's economy was near collapse and US intervention was necessary, President Clinton signed into law the 1994 Immigration and Nationality Technical Corrections Act, sec. 219 (108 US Statutes at Large [pt. 5]: 4305; Rothenberg 2000: 227). Its purpose was to regulate Mexican immigration by allowing farm labor to continue crossing the border, while at the same time preventing industrial workers from entering the United States. The Technical Corrections Act ended the 210A program, and in its place a new agricultural labor program was established that no longer allowed farmworkers to adjust their status. Under the 1994 reforms, agricultural workers were still welcomed, but only for temporary work. From that point on they entered under a new H-2A temporary visa program, and their numbers were allowed to increase annually. As mentioned in chapter 3, the annual quota in 1994 was 7,000; it increased to 41,000 by the late 1990s, and in 2005 rose to 75,000.

Once these agricultural labor provisions were set in place, President Clinton worked with Congress to enact policies to stop, dissuade, and discourage Mexicans from working in the United States. In late 1994, while Congress began proceedings to reform immigration policy and debate proceedings to deter undocumented migration, Clinton enacted border enforcement policies. His aim was to improve security along the US-Mexican border and reduce the movement of border crossers and contraband. To do so, Clinton increased the funding of the border patrol and established additional checkpoints on interior highways (Nevins 2002). On September 17, 1994, Operation Gatekeeper was funded to stop border crossers attempting to cross into California. It was designed to stop the movement of people who were unable to penetrate the Texas border. One year earlier, Operation Hold the Line had effectively reduced unauthorized entry through Texas. A few months after Operation Gatekeeper successfully stopped around 75 percent of border crossers from entering California, Mexicans shifted their route to Arizona, where crossing was simpler. Operation Safeguard was then instituted in Arizona, but it failed to accomplish its goal in comparison to the other two operations (Inda 2007: 145; Stephen 2007: xiv; Zavella 2011: 6). Basically, Mexicans were desperate to enter the United States, and as one entry point became difficult to penetrate, they traveled to other regions.

On September 30, 1996, the US Congress finally passed the Illegal Immigration Reform and Immigrant Responsibility Act (IIRIRA) (110 US Statutes at Large [pt. 4]: 3009). Its purpose was to discourage unauthorized immigration and to make it more difficult for US citizens and permanent legal residents to sponsor relatives who planned to immigrate to the United States. The sponsorship requirements mandated that family sponsors pay for all government benefits used by their relatives, with the exception of public schooling, emergency services, and soup kitchens.

Furthermore, under the act to discourage unauthorized immigration, specifically from Mexico, Congress increased the funding of border security along the US-Mexican border. One thousand border patrol agents were added, a border fence was to be constructed along parts of the California-Arizona border, new immigration detention centers were to be built, and the number of immigration inspectors was to increase in order to conduct more work site raids (ibid., 669).

The border enforcement initiatives were effective, but they did not stop undocumented immigration, because Mexico's economic recovery did not impact all citizens in the same manner. That is, although the IMF reported that in 1996 Mexico's economy was stable and foreign investments were safe, Mexico's poverty rate continued to remain over 60 percent throughout the 1990s (World Bank 2012). The Mexican government was also unable to produce the number of jobs needed, and citizens continued to live under austere public spending measures. Food assistance programs were nearly nonexistent, and farm aid to the small- and mid-scale agricultural sectors was not reinstituted (Moreno-Brid and Ros 2009).

US IMMIGRATION DEBATES:
A GUEST-WORKER PROGRAM

In 2001, after the attack of 9/11 upon US soil, the US stock market crashed, adversely affecting Mexico's economy. Mexico's GDP annual growth fell to 0 percent from 6.6 percent the previous year (IMF CR/6/352 2006). Many Mexican billionaires lost a large percentage of their assets, and foreign domestic investment in Mexico fell. Mexico, however, was not expected to enter another economic crisis, as its domestic economy was stable, with high demand for Mexican crude oil at an average price of US$18.61 per barrel ensuring a steady flow of cash (Pemex, *Statistical Yearbook 2011*). Furthermore, Mexico's economy was not expected to crash during this economic downturn, as Vicente Fox, who became president in 2000, had set in place domestic policies to assist the states and the private sector. Fox, unlike his predecessors, was not an economist; he was a businessman who shared the neoliberal philosophy that the private sector could lead the nation to prosperity. However, he also observed that the federal government's austerity plan caused productivity to stagnate. In complying with his presidential campaign slogan of "¡Ya Basta!" (That's enough!), Fox instituted reforms to increase social spending. He began by relaxing federal austerity policies imposed upon the states. To finance state programs, Fox allowed governors with solvent economies to borrow from international banks (IMF CR/4/418 2004). He also extended loans and grants for infrastructure projects to municipalities and state governments. Fox also reversed Salinas's and Zedillo's credit policies to the private sector. Rather than giving credit mainly to large-scale corporations, he extended credit to smaller firms, reserving only 22 percent for the largest corporations (INEGI 2012a: 243, 144). He increased the budget for education by 45 percent.

Fox's public sector reforms, however, were temporarily stalled when the US stock market crashed after 9/11. To deal with his inability to transform the quality of life for the working classes, especially in the rural sector, where unemployment was growing, Fox turned to the United States for assistance. He attempted to enact a guest-worker agreement for farm laborers in exchange for increasing oil exports to the United States (see chapter 5; Menchaca 2011). After extended discussions lasting for several years, the US Congress refused the request, even though there was strong evidence to indicate that US agricultural exports were a major contributor to farm labor unemployment. Between 1993 and 2005, 2.8 million agricultural jobs were lost (INEGI 2011: 221, 1996: 67) during a period when US exports of wheat increased by 400 percent and corn 200 percent (Seidband 2004: 24). National data also indicated that Mexico's agricultural exports were falling, adversely affecting

Mexican farmers. In 1980, Mexico's agricultural exports accounted for 14.7 percent of the nation's total exports, yet by 2006 agricultural exports had shrunk to 5.6 percent, and employment in agriculture fell by nearly 50 percent (Moreno-Brid and Ros 2009: 223, 228).[12]

The economic conditions in Mexico were not the US Congress's problem, however. Members of Congress therefore were not interested in negotiating a guest-worker program in exchange for an oil agreement. Data collected by the US Department of Labor offers insights as to why Congress was uninterested in negotiating a guest-worker program. According to the US Department of Labor, during the first years of the twenty-first century, farmers had an abundant supply of labor, and in some regions after the harvest season it was common for domestic workers to be underemployed (Levine 2004b). US farmers continued to depend on Mexican farm labor, but due to Mexico's inability to produce the needed jobs, Mexicans migrated north and created the ideal employment conditions for farmers. Employers did not need to compete for farm labor by offering higher wages. Throughout the country, farm labor wages failed to rise, and most farmworkers earned approximately half the wages of non-farm employees (Kandel 2008: 19; Levine 2009: 1). Even those guest workers who were employed by large-scale US corporations through the H-2A program experienced a drop in wages. During US president George W. Bush's administration, H-2A workers received US$1 to $2 less per hour due to the fall in prevailing wages as a result of an oversupply of labor (Bacon 2013: 251).

In 2000 the US Department of Labor estimated that American farmers employed from 1.8 million to 2 million farmworkers, of which over 80 percent were Mexican (Rothenberg 2000: 7, 40). The actual number of employees was probably much higher, since a large percentage of hired hands were undocumented and often paid off the books. Based on the National Agricultural Workers Survey, the undocumented farm labor population was estimated to be at least 55 percent of the total agricultural labor force (Levine 2004b: 2; NAWS 2005: 54). The actual percentage, however, was believed to be much larger, since surveyors thought it likely that not all those who reported they were legal had answered truthfully. The US Department of Agriculture also projected that the legal status of farmworkers was contingent upon their date of entry. It was unlikely that farmworkers who entered the United States prior to IRCA were undocumented, whereas those who entered during Mexico's mid-1990s economic crisis were generally undocumented. An estimated 86 percent of farmworkers who migrated to the US in 1996, and 98 percent of those who entered in 2001, were undocumented (Kandel 2008: 11).

Farm employment conditions began to change in 2005, with farmers throughout the nation reporting labor shortages. The US Congress failed to take action, however, as agribusiness labor shortages were part of a larger and more complicated immigration reform debate (see chapter 3). According to the Department of Homeland Security, the undocumented population in the United States had risen to 11.3 million, with over 58 percent concluded to be Mexican (Hoefer, Rytina, and Baker 2011: 2, 4). Of the undocumented immigrants, approximately 1.5 million were estimated to be farmworkers, with nearly 90 percent being Mexican. Congress therefore was unwilling to make any concessions on behalf of American farmers, as if it did, passing legislation to address the needs of other labor sectors would also be necessary. Congress also faced the difficult situation of assuring anti-immigrant advocates that if farmworkers were given amnesty, they would not move into industrial occupations. Unless a convincing argument was presented, the only method of addressing the needs of agribusiness was to pass comprehensive immigration reform.

In 2005 the interests of farmers conflicted with the general sentiment of Congress. Farmers asked for revisions of the H-2A guest-worker program so that they could more easily obtain labor. Their main demands were for Congress to increase the number of imported workers, and for employers not to be required to provide guest-worker housing. If Congress was unwilling to alter the program to meet these demands, then farm lobbyists requested that undocumented agricultural workers be allowed to adjust their status so that their illegality would not hamper them in moving from farm to farm (see chapter 3). Congress did not support the farmers' requests and defeated the AgJob bill of 2005. In fact, many members of the House of Representatives took a hostile position against Mexican undocumented people, including farm laborers, and passed House Resolution 4437, the "Border Protection, Antiterrorism, and Illegal Immigration Control Act of 2005." The resolution's aim was to deter unauthorized immigration by criminalizing the actions of Americans who rendered assistance to undocumented people. If American residents offered shelter, transportation, and medical assistance to undocumented aliens, they could be arrested and charged with a felony. Furthermore, penalties were raised for those entering the United States without authorization; the penalties now included a one-year prison sentence and prohibitions against adjusting their status in the future. The resolution did not have a guest-worker provision, and it contained policies to institute mass deportations.

During the Cancun Summit meetings from March 29 to 31, 2006, Mexican president Vicente Fox met with President Bush in Mexico. Fox presented Bush with an immigration proposal to dissuade him from supporting HR 4437 (Associated Press 2006; White House 2006). The House of Representatives had passed the resolution, but HR 4437 had stalled in the Senate. Democratic senators such as Hillary Clinton and Barack Obama were against the house bill and instead proposed passing comprehensive immigration reform that included giving amnesty to youths who were brought into the United States by their parents, increasing border enforcement funding, and passing a guest-worker program to fulfill labor shortages in diverse industries, including agriculture.

President Fox's proposal had been designed by Mexican academics, journalists, and social activists he had summoned to develop a plan to discourage the US Congress from passing HR 4437 (Castañeda 2007: 150). The plan contained a guest-worker program with recommendations to ensure that contract laborers returned. Furthermore, to encourage Mexicans to self-deport, the Mexican government would issue tax-deductible benefits to those returning, including deductions for building homes and businesses in their communities. Fox also informed President Bush that he was prepared to require all Mexicans entering the United States to present an exit permit; otherwise, immigration agents would not allow them to leave. This meant that Mexico would institute new surveillance techniques to prevent people from crossing the US-Mexican border without authorization.

As part of Fox's proposal, he offered Bush an oil agreement that would be favorable to US corporations. Fox was prepared to ask the Mexican Congress to increase the sale of oil to the United States at below global prices, and to authorize Pemex to purchase US$50 million of US drilling equipment (GAO 2007). President Bush was not impressed with the proposal, since he did not believe that Mexico had the capacity to increase its export sales to the United States (Castañeda 2007). Instead, Bush informed Fox that he would not intervene in the congressional debates. He also advised Fox that if Mexico was realistically prepared to negotiate a new oil agreement, it would be necessary to consider allowing foreign corporations to invest in Mexico's petroleum industry.[13] In the end, the summit meeting discussions over immigration, the guest-worker program, and enacting an oil agreement failed. To Mexico's advantage, however, the US Senate defeated HR 4437, and the criminalization and deportation policies were not instituted. The US Congress only supported increasing funding for border enforcement and building border fences along regions where undocumented border crossings were high.

Within Congress, fervent support to redress the problems of the agricultural industry continued, even though congressional representatives were divided on immigration policy. From 2006 to 2013, sixteen bills to stabilize the agricultural labor force were proposed, but they all failed to receive majority support.[14] The main policies addressed in the bills were the high cost of hiring farm labor, reforming the H-2A program, and extending amnesty to undocumented farmworkers. The bills fell under two main categories: AgJob Act or Comprehensive Immigration Reform Act. The primary purpose of the bills was to reduce the farmers' production costs, specifically for housing. In fourteen of the sixteen bills, granting undocumented farmworkers the opportunity to adjust their status to permanent legal residency was a key component directly related to reducing the farmers' housing expenses. Because most members of Congress did not support terminating current policy requiring employers to house guest workers, the adjustment process was developed to stabilize the farm labor force by no longer making this class of workers deportable. Since farmworkers under an adjustment process were not considered guest workers, employers were not required to house them. Moreover, the bills contained additional measures to reduce the farmers' housing expenses. Employers would be given the choice of either issuing their guest workers housing vouchers or making arrangements for their housing. The voucher system essentially placed the burden upon the workers to find inexpensive housing, and if the cost was beyond the voucher's value, that was their problem. Most of the bills also proposed ending the employer's responsibility to finance a worker's round-trip transportation from Mexico to the United States.

Although the housing and transportation proposals shifted the living-expense costs to the individuals who were the least able to afford them, the amnesty provision was a humane policy and demonstrated that the US Congress considered farm laborers an essential and strategic component in lowering the cost of food production. Furthermore, supporters of the resolutions had also proposed that the families of farmworkers be allowed to adjust their status to permanent legal residency if the farmworkers applying for adjustment agreed to remain employed in farm labor for a period of three to five years after adjustment. This generous adjustment policy, however, must be seen as a strategic rather than an altruistic action, since it has been well documented that some of the children in farmworker families traditionally adopt their parents' occupation.

To the present day, the US Congress has been reluctant to pass legislation particular to farm labor, as critics on both sides of the debate propose that comprehensive immigration reform must be addressed first. Opponents who are against any type of amnesty proposal argue that the border must be secured before discussions on farm labor can proceed. Those who favor giving amnesty to undocumented people also oppose debating only farm issues, because afterward Congress will not want to negotiate amnesty for the general public. When IRCA was passed, in 1986, the needs of the agribusiness industry became the turning point in leading Congress to compromise. Thus, if Congress first agrees to revise the guest-worker program and give amnesty only to farmworkers, those who support a general nonoccupational amnesty mandate would lose their negotiating leverage.

One point on which most members of Congress agree is that Mexican farm labor is a valuable and scarce commodity in the United States. The point of disagreement, however, is whether Congress should negotiate a new accord with Mexico or leave matters as they are, meaning allow a supply-and-demand formula to regulate the farm labor market. That is, if Congress does not take action, then Mexican labor will continue to flow, because farmers need labor and Mexico cannot produce the necessary jobs to employ its people. The consequence of inaction is that Congress indirectly encourages farmers to continue recruiting and employing undocumented workers, leading to unplanned immigration.

GRIDLOCK: US CONGRESS FAILS
TO PASS IMMIGRATION REFORM IN 2013

In 2013, Republicans and Democrats nearly reached a compromise on immigration reform. Republicans overwhelmingly supported policies to address the problems faced by the agricultural sector, while Democrats favored comprehensive immigration reform. Both sides supported increasing border security and developing a national electronic monitoring system to verify who was eligible to work in the United States. By late June 2013, it appeared as if the US Congress might reach an agreement, but six months later the momentum to reform immigration policy slowed down. The upcoming midterm election of November 2014 dissuaded congressmen from taking a stand, as they were reluctant to anger potential voters who did not support their position. This gridlock propelled a series of counteractions that ultimately moved immigration reform into the realm of the executive and judicial branches. The events leading to the gridlock began two years earlier, when US president Barack Obama warned Congress that it was critical they pass immigration legislation.

In 2012, the large size of the undocumented population within the United States and the needs of the agricultural industry pushed President Obama to pressure members of Congress to begin debates on reforming immigration law. When neither Democrats nor Republicans were prepared to put forth a congressional bill addressing immigration policy, Obama intervened and on June 15 asked Janet Napolitano, secretary of homeland security, to implement an executive prosecutorial discretion order (US Department of Homeland Security 2012). This was done as a state of emergency, because the president professed that congressional inaction had culminated in a system that was inefficient, unfair, and broken (O'Brien 2012). According to Obama, Congress needed to pass legislation to stop employers from hiring undocumented workers and honor the US heritage as a nation of laws and a nation of immigrants. A fair system of legal immigration must be instituted to ensure that immigration law no longer hindered the nation's economy. Obama's foremost concern was to focus enforcement resources on the removal of individuals who posed a national security or public safety risk, including repeat immigration law offenders and immigrants convicted of violent crimes or other felonies. The president's plan also sought to safeguard against the deportation of undocumented youth to countries where they may never have lived and could not speak the language. Presidential discretion orders, which are used in many other areas, were especially justified here because of congressional inaction. By passing the "Prosecutorial Discretion with Respect to Individuals Who Came to the United States as Children," which came to be known as DACA (Deferred Action for Childhood Arrivals), youth who had entered the United States before age sixteen and had graduated from high school or were in the process of graduating would not be deported. Those who had graduated must not have turned age thirty-one as of June 15, 2012, or be attending college or in the military. Those who applied and obtained provisional status were labeled "Dreamers" and were given temporary deferred deportation protection as well as work permits.

President Obama's prosecutorial discretion order angered many in Congress and convinced Democrats and Republicans to form a bipartisan committee to address immigration reform. His directive certainly triggered immediate action, as many in Congress projected that further presidential discretion orders were forthcoming if they did not introduce a plan. The need for action was especially urgent for the agricultural industry, as it was public knowledge that Obama did not support increasing the annual number of imported agricultural guest workers, nor did he support an adjustment policy that pertained only to undocumented farmworkers.

To expedite negotiations within the bipartisan committee, President Obama unveiled his comprehensive immigration proposal during the State of the Union address on January 29, 2013. Called "Creating an Immigration System for the 21st Century" (White House 2013), the announcement led the bipartisan committee to accelerate the time frame for their plan (US Senate Judiciary Committee 2013a). Two months later, on April 17, the Senate Judiciary Committee unveiled its immigration plan, and a few days later it held a hearing to determine if the bill would move to the Senate floor for open debate (US Senate Judiciary Committee 2013b). The hearing began on April

23, and the plan was introduced as Senate Act 744, The Border Security, Economic Opportunity, and Immigration Modernization Act, of 2013.

To move discussion from the committee hearing room to the Senate floor, it was critically important for representatives of the Department of Homeland Security (DHS) to certify that advances had been made in securing the border; otherwise, debate would not begin in the Senate. DHS Secretary Napolitano testified that the border was 82 percent secure. She reported, however, that improvements along the border were needed. To better detect border crossers, she advised flying drones from the Texas border fifty miles into the state's interior. Napolitano also requested $1.5 billion in additional funds to install new technology and complete the fence along the US-Mexican border. In addition, the Texas border surveillance budget needed to be increased by $3 billion.

To assure Congress that with this additional funding the DHS could meet the Senate's mandate to secure the border, Secretary Napolitano outlined DHS's current achievements. As a result of border patrol apprehensions, the flow of border crossers had decreased in the previous four years by 49 percent. Furthermore, she testified that the DHS had also improved other aspects of border security. In 2013 the DHS had seized 71 percent more currency than in 2009, 39 percent more drugs, and 189 percent more weapons along the southwestern border. This had led to a significant decline in crime in New Mexico, Arizona, California, and Texas (US Department of Homeland Security 2013).

After the hearings, US secretary of agriculture Tom Vilsack urged Congress to compromise and pass the Senate's act, as farmers were experiencing severe problems due to Congress's refusal to debate issues pertaining to farm labor. In a meeting of the American Royal Center in Kansas City on June 21, 2013, Vilsack urged Congress to pass Senate Act 744. He believed that many crops in the United States were not harvested because farmers did not have a stable and secure labor force. The secretary stated, "Today, sadly, we've got crops that are growing in this country that are not being harvested, we've got dairy producers that can't find an adequate workforce" (Alonso 2013; Hanchard 2013). Vilsack also stated that most farmworkers were undocumented: "The reality is that many—perhaps most—of the US farmworkers are undocumented, and it has been like that for quite some time. We have an immigration system which is completely broken. I believe that undoubtedly this is the case" (Univision 2013).

On June 27, 2013, the Senate debated the bipartisan act and voted 68 to 32 in favor of the measure (US Senate Act 744). Many Republicans who favored only immigration reforms that would solve agribusiness problems chose to end the gridlock by voting in support of the act, since it was the only way of expediting agricultural policy. Senate Act 744 contained many favorable policies lowering the farmers' labor costs.

The bill then moved on to the house for consideration. Senate Act 744 rested on four pillars: (1) create a tough but fair path to citizenship for unauthorized immigrants currently living in the United States that is contingent upon securing the borders; (2) reform the legal immigration system to issue visas for those in the occupations and with the social characteristics needed to build a stronger economic system in the United States; (3) create an effective employment verification system; and (4) establish an improved process for admitting future workers to serve the nation's workforce needs.

Within the Senate plan, it was apparent that the needs of the agribusiness industry were a priority commanding immediate action. These provisions echoed the same policies contained in earlier versions of the AgJob acts. The Senate supported establishing an expedited and separate process of adjustment for agricultural workers and their families. Agricultural workers and their dependents would receive blue-card visas. The legal status of a farmworker family would remain provisional until the individual who qualified for adjustment completed the occupational residency requirement of approximately eight years (ibid., sec. 245F). This policy was to ensure that employers would not be fined for hiring undocumented labor. It also relieved employers from paying housing expenses, since those eligible for adjustment did not qualify for guest-worker housing or vouchers.

The process was certainly more streamlined for farmworkers than for other immigrants seeking legal adjustment. Only those youths who had applied for deferred action were given a similar expedited path (ibid., sec. 245c[c]). Under President Obama's 2012 directive, Dreamers obtained the legal status of Registered Provisional Immigrants (RPI), which distinguished them from other undocumented people by giving them temporary legal residential status. Dreamers were treated as subjects eligible for expedited adjustment because it was their parents who had broken the law. The termination of their provisional status, however, would be determined on a case-by-case basis, but could not be less than five years (ibid., secs. 2103 and 2450; US Department of Homeland Security 2012). In the case of other undocumented applicants, the adjustment process was lengthy and could take more than eighteen years to complete. Under Senate Act 744, undocumented applicants would remain in provisional status until all eligible applicants who had not entered illegally had received a permanent resident visa.

The Senate plan also contained a very liberal guest-worker program provision allowing 112,333 guest workers to enter each year under contracts lasting as long as three years and renewable if the employee was in demand (US Senate Act 744: sec. 218a). The housing voucher system debated in earlier AgJob bills was also included in the Senate act. Farmers were given the option of providing housing or issuing a housing voucher. However, to avoid placing the burden on local communities, the act mandated that housing vouchers could be issued only when state governors certified that a farming community had sufficient migrant rental units; otherwise, farmers would not be able to issue housing vouchers (ibid.).

Opponents of Senate Act 744 introduced an alternate plan that did not contain comprehensive immigration policies and addressed only the problems of the agricultural industry. The bill did not have as much support as the Senate Act, but it nonetheless aired the political position of those who were against any type of amnesty. House Resolution 1773, titled "Agricultural Workers Act," did not support any type of permanent adjustment provision, but it did create a new semipermanent category for agricultural workers. Unauthorized workers would be eligible to adjust their status to temporary guest workers (US House Resolution 1773: sec. 218A, 3P; sec. 218B, sec. 7). They would be given a three-year renewable contract, with a stipulation that upon renewal they had to return to their home country for a short period of time. No limitation on the renewal contract was set. This bill also favored establishing a massive guest-worker program, much larger than the ones that had existed during the bracero program. The bill proposed importing 500,000 agricultural workers each year, more than four times the number requested in Senate Act 744, thus indicating that the Senate's labor shortage estimates were much larger than projected. The bill also contained other mean-spirited policies that Mexico most likely would not have accepted if a guest-worker program were to be renegotiated. Guest workers would not receive a contracted set hourly wage, and they would be required to pay for their round-trip travel and farm labor housing (American Farm Bureau Federation 2015).

By December 2013, neither bill had been introduced for debate in the House of Representatives. Sponsors of the bills projected that the debate would resume after the November 2014 midterm election, when the representatives no longer feared voter retaliation. Senate Act 744 was expected to be debated in Congress immediately after the elections. If it failed, then immigration reform would be debated on a piecemeal basis.

IMMIGRATION DEBATE ENTERS THE COURTROOMS

On November 4, 2014, the Republican Party won the midterm elections and took control of both houses in Congress. With a Republican majority in Congress, it was no longer necessary for those Republicans who opposed amnesty to compromise. The news media immediately projected that Senate Act 744 would fail if the amnesty provision for the general unauthorized population was not removed. Nonetheless, President Obama urged Congress to submit an immigration bill and try to resolve some of the issues.

On November 20, 2014, President Obama announced that due to the failure of Congress to take action on immigration policy, he would need to enact another executive directive to protect the US economy. At that time approximately 11.4 million people were estimated to be living unauthorized in the United States, 59 percent of whom were from Mexico (Baker and Rytina 2013). According to the president, the precarious legal status of the undocumented population stalled the economy, allowed employers to defy employment laws, and prevented the full taxation of these people.

For altruistic reasons, President Obama's directive was not enacted. Television and newspaper reports proposed that his concerns over congressional inaction were also highly motivated by political pressure from the Latino community and other pro-amnesty activists. These constituencies continued to demand that he "make good" on his 2008 presidential campaign promise not to deport productive and lawful undocumented immigrants (Oleaga 2015a). On November 20, 2014, Obama ordered the secretary of Homeland Security to implement Deferred Action for Parental Accountability (DAPA) and revise DACA (National Immigration Law Center 2015). DAPA suspended the deportation of undocumented parents who had entered the United States no later than January 1, 2010, and had children who were US citizens, were permanent legal residents, or qualified for DACA. The parents also must have no convictions for felony crimes. This was only a temporary measure to address the controversy over which types of undocumented immigrants should not be deported, made during a period when Congress was gridlocked over immigration reform. DAPA would give applicants a three-year work permit and allow them to stay without fear of deportation. The permit was subject to renewal on a case-by-case basis.

DACA was also revised, and the age limit previously placed on applicants was removed. The revision allowed applicants age thirty-one and over to be eligible for deferred action. This was also a temporary measure that could be nullified or altered by Congress. Neither DAPA nor DACA guarantees an applicant permanent legal status or places them on the path toward citizenship. Only Congress has the legal power to enact these actions.

President Obama expected that his most recent executive directive would put pressure on Congress to take action. Instead, the Republican majority leaders of the House and Senate announced that Obama's action was illegal and irresponsible, and not only would delay but in fact had temporarily ended any bipartisan debate (MSNBC 2014). They also promised to reverse DAPA and DACA by defunding the president's deferred action directives (Shabad and Marcos 2014). On December 2014, debate over HR 1773 and Senate Act 744 was to take place in the House of Representatives, but House leader John Boehner postponed the debate indefinitely. Both bills died.

Processing of applications for DAPA and the revised DACA program was to begin in February 2015. However, Texas attorney general Greg Abbott and twenty-five representatives of other states submitted a petition to the US Southern District Court of Brownsville, Texas, to prevent the processing of the applications, and the judge, Andrew Hanen, agreed with them. An injunction was executed by the court (*State of Texas v. United States of America* 2015). The representatives of the dissenting states argued that President Obama did not have the authority to issue deferred action orders, because only Congress has the authority to implement immigration law. Judge Hanen ordered a temporary halt of the programs, but also opined that Obama had overreached his legal authority.

The injunction did not reverse President Obama's executive discretion orders, but it did halt the processing of applications. Only the higher courts could determine whether the president's executive orders were lawful or a violation of the constitutional division of power between Congress and the executive branch. The US Department of Justice immediately submitted an appeal against Judge Hanen's orders, asking the Fifth Circuit Court of Appeals of Louisiana to lift the temporary injunction and allow the processing of the applications (*State of Texas, et al. v. United States* 2015). Amicus briefs of sixteen states were introduced in support of the Department of Justice. On May 29, 2015, the Obama administration experienced a setback when the Fifth Circuit Court chose not to reverse the injunction on the grounds that the Justice Department had not proved that the delay of the programs was an emergency or hindered the nation (Oleaga 2015b; Tanfani 2015). The three-judge panel hearing the appeal was divided, issuing a two-to-one vote not to lift the injunction. Two judges opined that the defendants (i.e., the states' attorneys) had offered compelling arguments demonstrating that their states would suffer if President Obama's orders were implemented before the courts deliberated on the legality of his actions.

On the day of the ruling, news reporters projected that the Justice Department would appeal for an emergency hearing before the US Supreme Court (Gomez and Jackson 2015). Instead, the department announced that it would no longer appeal the injunction. It would instead focus on implementing the programs and preparing the oral arguments on the president's legal authority to take action without the approval of Congress. On July 10, 2015, the Fifth

Circuit Court, in a divided decision on a two-to-one vote, ruled that the injunction against DAPA and DACA would remain in force and that the states had the right to challenge the validity of President Obama's prosecutorial discretion orders. The same day the Department of Justice announced that an appeal would be submitted to the US Supreme Court (National Immigration Law Center 2015).

IMPLICATIONS OF CONGRESSIONAL GRIDLOCK
FOR FARM LABOR RECRUITMENT

The implications of the suit against the executive branch's deferred action, as well as the reluctance of the US Congress to debate immigration reform, stall any change on immigration policy related to farm labor. The most important policies that will be postponed are deciding whether undocumented farmworkers will receive expedited amnesty, renegotiating a substantial expansion of the current guest-worker program with Mexico, and decreasing the cost of housing for employers by establishing a voucher system. In the meantime, Congress has authorized the US Department of Labor to increase the number of agricultural guest workers. By mid-2015, the US Department of Labor had certified the entry of 83,579 guest workers, more than the annual limit of 75,000 that was common in the previous five years (USDA 2015b). Farm labor wages have also increased throughout the country to an average of $10.80 an hour, but they range from $9.50 to $12.33.

Wages vary across the country, however, as does the type of labor force used. Employers in the US border states report paying lower wages and generally employing local labor to meet demand rather than importing guest workers. Severe labor shortages are unusual, but they do occasionally occur. The US Department of Agriculture estimates that the majority of workers in the border states are local residents and the undocumented. An inverse situation, however, has occurred outside the border region. The southern and northern states have reported ongoing labor shortages and the need to increase wages to compete for available labor. In 2015 the states reporting the most severe labor shortages and increased reliance on guest workers were North Carolina, Louisiana, Georgia, Kentucky, New York, and Washington. The US Department of Agriculture reports that there are 1.1 million farmworkers legally employed in the United States, but it estimates that the farm labor population is much larger. On the basis of the findings of the 2012 US Agricultural Census, the labor force is estimated to be approximately 2.7 million (ibid.; US Census of Agriculture 2015).

In sum, whether or not Congress remains gridlocked on legislation to reform immigration policy, US farmers will need to continue employing agricultural labor. The current H-2A guest-worker program may continue to regulate a small sector of the farm labor force, yet a large percentage remains unregulated. How this affects the lives of employers and employees is uncertain, as both groups are breaking the law and their actions are not under the scrutiny of US agencies.

CHAPTER FIVE

Mexico Reopens the Oil
Industry to US Investors

*C*HAPTER 4 EXAMINED MEXICO'S ECONOMY FROM THE 1980s to 2013, focusing on the effects Mexico's financial crises have had on the agricultural industries of the United States and Mexico. This chapter will revisit the economic events that transpired in those historical intervals and examine how the financial crises shaped US-Mexican oil trade relations. The US government will be shown to have yielded considerable influence over Mexico's oil industry during the crisis periods by demanding oil exchange conditionality agreements when extending loans to Mexico. The oil agreements greatly benefited US consumers, but at the same time they intensified US dependence on Mexican oil. In 2015, Mexico continued to be the third-largest crude oil supplier to the United States (EIA 2013b, 2015b).

Its growing dependency on Mexican oil has made the US government very attentive to how much oil Mexico produces and the price it sets for a barrel of oil (see US Senate Foreign Relations Committee 2006: 25). In this chapter, it is argued that for the US government, it is now vitally important that Mexico not only price its oil reasonably but also meet the US consumption demand. Officials of the Mexican government are aware of this supply-and-demand political relationship and acknowledge that if Pemex is unable to fulfill demand, Mexico may lose its strategic international standing. The Mexican government therefore has been pressured periodically to increase its oil production level at the cost of depleting its reserves. Meeting US demand has been difficult for Mexico, and by 2013 the Mexican government had decided to allow US investors to reenter the industry and assist Mexico in improving productivity.

The economic crashes of the 1980s and 1995 will be examined as turning points when the US government became involved in shaping Mexican oil policies. The US-Mexican loan accord of 1990 will also be explored, because although it was not prompted by a crisis, it demonstrated that the Mexican government needed to lower its interest loan rates by negotiating an oil exchange agreement advantageous to US national security. This chapter will conclude with a discussion of Mexico's energy reforms of 2008 and 2013, which gradually but radically altered Articles 27 and 28 of the Mexican Constitution, which now allow foreign investments in the Mexican oil industry.

US NATIONAL SECURITY: BUILDING US STRATEGIC PETROLEUM RESERVES

American consumption of crude oil began in the 1890s, when the popularity of automobiles created a great demand for gasoline. With the growing use of automobiles, gasoline derived from crude oil emerged as the preeminent source of energy. By 1954, crude oil had many uses and replaced coal as the main energy source in the United States (EIA 2013a; Shao 1956: 2). The automobile had become an important part of American culture, with families depending on cars in carrying out their daily activities. Railroad companies also switched to crude oil as their main energy source, and Americans began to rely on fuels derived from crude oil to heat their homes and cook their meals. Since then, US residents have been the largest consumers of crude oil in the world, currently consuming 21 percent of the world's total crude oil–derived products (EIA 2013a, 2015a).[1]

In 1973, the first oil embargo launched against the United States by Middle Eastern nations led the US government to establish a strategic petroleum oil reserve to defend national security. Americans needed to be protected from future oil embargos provoked by political conflicts against the United States. The conflict had begun after the US government rendered economic aid to Israel during its invasion of territories in Syria and Egypt. Part of the economic aid rendered to Israel by the US government consisted of shipments of petroleum supplies, since Arabic nations had stopped selling oil to Israel. In retaliation against US involvement in the Middle East, the Organization of Arab Petroleum Exporting Countries (OAPEC) and Arab members of OPEC boycotted the shipment of oil to the United States. The conflict finally ended in 1974, after the United States negotiated Israel's withdrawal from the Sinai region (Zalloum 2007).[2] This global conflict convinced the US government that it must quickly establish a massive stockpile of crude oil, and for that purpose it founded the Strategic Petroleum Reserve (SPR) in 1973 (see Gonzalez, Smilor, and Darmstadter 1985; US Department of Energy 2014).

A few years later, when crude oil supplies from the Middle East were once again cut off following the Iranian hostage crisis of 1979, the US government learned that it was necessary to accelerate the storage of crude oil if the United States was to have enough oil to endure an embargo. During the Iranian conflict, crude oil supplies from the Middle East abruptly declined, and it became necessary for the US government to find alternate sources of oil. As discussed in chapter 4, Mexico supplied the United States with 40 percent of its crude oil during this crisis.

Following the Iranian hostage conflict, Mexico emerged as an important player in international petroleum politics. Mexico's ability to supply the United States with large quantities of crude oil shipments proved that Pemex's exports could destabilize an oil embargo. Likewise, Mexico's ability to quickly supply an order indicated that Pemex shipments could potentially affect the global price of crude oil.

In 1979, OPEC attempted to convince leading oil-exporting nations that were not members of OPEC to form a network to regulate the price of crude oil. OPEC's aim was to partner with oil-producing nations and have greater control over the price of oil. This was critical to OPEC nations, as they produced 50 percent of the world's oil but found it difficult to set higher prices because nonmember nations sold their oil at lower prices (Al-Chalabi 1984: 58). If a consortium of nations agreed upon the amount of oil supplied globally, then large-scale-consuming nations such as the United States and Great Britain would be unable to manipulate the price of crude oil. British Petroleum, Royal Dutch/Shell, Exxon, Mobil, Texaco, and the Gulf Corporation had been able to control the price of oil by boycotting OPEC if the price was inadequate. Petroleum corporations could hold off from purchasing oil for long periods, since they owned large crude oil deposits around the world or had licenses to drill in federally owned lands. They also controlled the price of oil by conducting business with nations that did not support OPEC policies. US and British oil corporations often sought assistance from the US government by asking to drill in federally owned land. The US government owned massive deposits in the deep water of the Gulf of Mexico, the Rocky Mountains, Alaska, and the US Gulf coast. Over the years, US crude oil deposits ranged from eighth- to eighteenth-largest in the world (GAO 2008). For the US government, leasing federal lands during politically tense moments was advantageous, since Americans consumed the largest percentage of the oil in the market. Prices could be lowered or at least prevented from rising by allowing corporations to drill in federal lands (Gonzalez 1985).

This scenario was not ideal for OPEC, since the US government could release large amounts of oil to affect the global price of oil. However, if OPEC could convince independent oil-exporting countries such as Mexico to join its network, then US-British coalition would be less effective (Kubbah 1974: 7; Pemex, *Statistical Yearbook 1977*: 54). By 1979, after the Cantarell discovery in the Gulf of Mexico, the Mexican government's role in global oil politics had become more significant. Mexico's proven reserves had moved from the fourteenth- to the fifth-largest holding in the world (see table 5.1; Pemex, *Statistical Yearbook 1988*: 6–8). Mexico's proven reserves now exceeded US proven deposits (see table 5.2).[3] The two nations together owned over 62 percent of the proven crude oil deposits in North America (Pemex, *Statistical Yearbook 1977*: 51; Pemex, *Statistical Yearbook 1988*: 6–8).

For OPEC it was critical that Mexico become a member, or at least agree not to interfere. But in 1979, Mexican president Jose López Portillo declined to join OPEC, because it was not in the nation's best interests to become a minor partner in a network composed of the largest oil-producing nations (Díaz Serrano 1989). Essentially, Mexico would lose its independence to an organization controlled by Middle Eastern special interests (Meyer, Sherman, and Deeds 2007). For example, in 1979 Mexico's proven reserves had risen to an impressive 33.4 billion barrels, but in comparison, Saudi Arabia's proven crude oil reserves were 163.4 billion barrels, Kuwait's 65.3 billion, and Iran's 58 billion (Pemex, *Statistical Yearbook 1988*: 6–8).

Most other non-OPEC nations refused to join as well. In early 1981, OPEC retaliated by releasing millions of barrels of heavy crude oil, while at the same time reducing the sale of light crude oil. Light crude oil was the preferred mix in the global market. Since most non-OPEC nations produced heavy crude oil, this immediately created an oversupply that lowered the price of heavy crude and caused the profits of nonmember nations to fall. OPEC nations were also adversely affected, but they made up for the lost revenue by increasing the price of light crude oil. Mexico's heavy crude oil prices fell, but since Mexico also produced light crude oil, it could offset any economic losses by raising the price of its Isthmus light. OPEC expected its actions to serve as a warning to nonmember nations, but it knew that Mexico could withstand the shock. Mexico was expected to reconsider its policies and follow OPEC's lead by raising the price of its light crude oil and reducing production.

Mexico once again refused to follow OPEC, however, since its pricing practice went against US interests. In the meantime, the US Congress informed its oil exporters, including Mexico, that if oil shipments were reduced or prices increased, their duty-free exports would be immediately suspended and trade agreements opened to renegotiation (Randall 1989: 18). Notwithstanding the US government's warning, Jorge Díaz Serrano, the director of Pemex, advised President López Portillo to work with OPEC in pricing light crude oil. The president refused, since he was not about to lose US commerce agreements due to OPEC's price war. The US government and US corporations were Mexico's trading partners, and he was not prepared to alienate them. Because US corporations continued to purchase 66 percent

TABLE 5.1. INTERNATIONAL COMPARISON: MEXICO'S PROVEN CRUDE OIL RESERVES AND PRODUCTION LEVEL RANKINGS, SELECTED YEARS

Year	Proven Reserves	Production	Historical Event
1920		2	
1930		6	
1950		9	
1960	13	10	
1970	17	15	
1975	19	16	
1976	15	15	Cantarell Gulf discovery
1977	14	15	
1978	8	13	Cantarell Gulf deposits concluded to be massive
1979	5	15	
1980	5	7	
1981	5	4	Mexico's economic crisis begins
1982	4	4	
1983	4	4	
1984	4	4	
1985	4	4	
1986	4	6	Mexico joins GATT
1987	8	5	Black Monday
1988	8	6	
1989	8	7	
1990	8	6	Oil Exchange Agreement
1992	8	6	
1993	8	6	
1994	8	6	
1995	8	7	Recovery plan negotiations
1996	8		
1997	8	7	Reserve recalculation discussions
1998	9	7	
1999	9	7	
2000	9	8	
2001	9[1]	7	
2002	9	6	Reserve recalculation discussions
2003	13	6	Cantarell Gulf oil production decline
2004	14	6	
2005	14	6	
2006	14	6	
2007	16	6	

TABLE 5.1. CONTINUED

Year	Proven Reserves	Production	Historical Event
2008	17	6	Mexico's privatization debate
2009	17	6	
2010	17	7	
2011	17	7	
2012	18[2]	10	
2013	18	10	Mexican Constitution revised
2014	17	10	

Sources: Pemex, *Statistical Yearbook 1977* (International Comparison section) for 1920–1950; *Statistical Yearbook 1988* for 1960–1988; the rest for corresponding years.

[1]In 2002 the Securities and Exchange Commission recalculated Mexico's proven reserves for 2001. This changed Mexico's international rating to 13 from 9 (Pemex, *Statistical Yearbook 2003*: 55). OPEC rates Mexico at 10 (*OPEC Annual Statistical Bulletin 2001*: 34).

[2]In 2012 Pemex rated Mexico to be 17th in proven reserves, but in 2013 revised the estimate to 18th.

of Mexico's crude oil, López Portillo feared that if Pemex export prices were raised, the US government would retaliate (Díaz Serrano 1989; Pemex, *Statistical Yearbook 1990*: 38). López Portillo preferred to maintain the status quo, since US corporations were doing business as usual and continued to purchase Mexico's more expensive heavy crude oil even though OPEC was offering lower prices. The previous year, Mexico's annual crude oil exports had skyrocketed to 303 million barrels, with US clients as the main importers (Pemex, *Statistical Yearbook 1988*: 121).

The president was also concerned that if Mexico worked with OPEC, the US government would advise US banks to raise their interest rates when rescheduling Mexico's loan payments (Al-Chalabi 1984; Boughton 2001). Following the discovery of oil in the Gulf of Mexico, the Mexican government was heavily indebted to the US Treasury and to American corporations. In a short time the Mexican government had nearly doubled its external debt, borrowing heavily from US banks in expectation that Mexico's oil production and exports would triple and increase the federal revenue (*Congressional Record 1982b*: 23438; *Congressional Record* 1982c: 22463). Mexico's external debt rose from US$26.1 billion in 1976 to $50.7 billion in 1980 (*Congressional Record* 1982a: 24019; Martínez Fernández 1996: 18).

After Mexico failed to follow OPEC's price system, OPEC members lowered their price of light crude in the winter of 1981, placing Mexico in a

TABLE 5.2. US ESTIMATES: MEXICO AND UNITED STATES PROVEN
CRUDE OIL RESERVES, 1980–2014 (IN BILLION BARRELS)

Date	Mexico[1]	United States	
1980	31.250	29.810	
1981	44.000	29.805	
1982	56.990	29.426	
1983	48.300	27.858	
1984	48.000	27.735	
1985	48.600	28.600	
1986	49.300	28.416	
1987	54.653	26.889	
1988	48.610	27.256	
1989	54.110	26.825	
1990	56.365	26.501	
1991	51.983	26.254	
1992	51.298	24.684	
1993	51.298	23.745	
1994	50.925	22.957	
1995	50.776	22.457	
1996	49.775	22.351	
1997	48.796	22.017	Recalculation discussions
1998	40.000	22.546	
1999	47.822	21.034	
2000	28.399	21.765	

financially perilous situation. Díaz Serrano immediately lowered the price of Mexican light by US$4 to compete with global prices. President López Portillo objected to the director's preemptive action, as he did not believe that sales would plummet if Pemex failed to lower its prices (Díaz Serrano 1989: 107). The president ignored Díaz Serrano's recommendations and raised the price of light crude oil by $2, expecting US petroleum corporations to support Mexico by purchasing the more expensive Mexican blend. This was logical, since he had chosen not to form a compact with OPEC (Goodsell 1981: 1). To President López Portillo's disbelief, however, US corporations instead purchased the less-expensive oil. By April 1982, Mexican crude oil sales had plummeted and exports had fallen from 1.3 million barrels a day to 300,000 (Randall 1989: 174). This devastated the Mexican economy, and within six months the federal government had insufficient capital to finance its public institutions, give assistance to the states, or pay its foreign debt (IMF EBD/82/99 1982).

TABLE 5.2. CONTINUED

Date	Mexico[1]	United States	
2001	28.260[2]	22.045	
2002	26.941	22.446	Recalculation discussions
2003	12.622	22.677	
2004	15.674	21.891	
2005	14.600	21.371	
2006	12.882	21.757	
2007	12.352	20.972	
2008	11.650	21.317	
2009	10.501	19.121	
2010	10.404	20.682	
2011	10.420	23.267	
2012	10.359	26.544	
2013	10.264	36.500	
2014	10.073[3]	33.400	

Sources: EIA 2013c; EIA 2014c; EIA 2014d.

[1]See appendix A for Pemex data. US data used to maintain comprehensive history in one source. Pemex data offers parallel analysis.

[2]In 2002, Mexico's proven reserves were recalculated for 2000 and 2001 (Pemex, *Statistical Yearbook 2003*: 55). In 2002, Mexico rescheduled its long-term loans with international banks (Pemex, *Statistical Yearbook 2002*: Financial Review Appendix).

[3]Pemex estimates are much larger than the EIA's. Mexico estimates 13.438.5 (Pemex 2014d).

The US Congress became concerned with Mexico's economic setback, but it was not prepared to encourage US corporations to buy Mexican oil at inflated world prices. Instead the US Treasury was prepared to assist Mexico by lending it money, as it was common knowledge that Mexico's oil revenues had fallen by 70 percent and its government would be unable to make its quarterly foreign debt payments to US banks without financial assistance (*Congressional Record* 1982c: 22463). At that time, 48 percent of the Mexican federal budget came from oil revenues (Manzo Yépez 1996: 53).

After prolonged negotiations between representatives of the Mexican government, the US Treasury, and the IMF, a financial plan was completed in August 1982. It was composed of various loan arrangements, with most of the credit being issued by the US Treasury, US banks, and the IMF (Boughton 2001: 292, 311). Mexico was to receive a total loan amount of US$12 billion. As part of the US bailout package, Mexico agreed to supply the US Treasury with US$1 billion in crude oil as a form of prepayment. The oil exchange agreement with the treasury had been strongly disputed by President López Portillo, as he believed Mexico was being forced to sell its oil at below-market prices. He considered the US actions unethical, since Mexico's predicament had been largely caused by the president's refusal to side with OPEC. However, with the Mexican economy in total collapse, López Portillo conceded to US demands. The US Treasury agreed to pay Mexico US$27.40 a barrel for crude oil, when global prices were $32 (Randall 1989: 181). The oil purchased was shipped to the US Strategic Petroleum Reserve (*Congressional Record* 1982c: 22463).

As the price of crude oil continued to fall, Mexico's economy failed to recover. This was devastating for Mexico, but for US petroleum corporations, Mexico's political blunder resulted in an inverse gain. As the price of oil continued to fall from US$26.42 a barrel in 1983 to $11.86 in 1986, corporations were able to purchase inexpensive oil (see table 4.1). The US government also profited, since it continued to purchase large volumes of crude oil for its Strategic Petroleum Reserve. By 1986, 40 percent of the US SPR was filled with Mexican crude oil and stored in underground salt caverns along the Gulf of Mexico (EIA 2013a; EIA 2013d; Randall 1989: 181; US Department of Energy 2014).[4] Two years later Mexico continued to export further shipments to the United States, with 90 percent directly used to fill the SPR. In 1988 the US Strategic Petroleum Reserve had the largest crude oil stockpile in the world, amounting to 540 million barrels.

Although Mexico had been a willing partner in its trade accords with the United States, in 1989 it sought to gain a higher price for its oil and therefore formed a closer relationship with OPEC. This was politically bold for Mexico, since the federal government continued to rely on US financial assistance. Mexico had also begun to develop a similar relationship with Japan. Japan had become Mexico's third-largest importer of crude oil and also was a significant creditor (Pemex, *Statistical Yearbook 1990*: 38). Essentially, Mexico attempted to separate its crude oil export business from its economic relations with the US government. Developing a competitive export market infrastructure was necessary if Mexico was to gain some independence from its main creditor. If it did not, then in the future the US government would be able to determine Mexico's oil prices in return for financial assistance. By 1989, Mexico had borrowed a series of loans that raised its external foreign debt to US$95.1 billion (IMF SM/94/41 1994: 59). US creditors continued to be Mexico's main lenders. It was the right time to enact a business arrangement with OPEC, since Mexico expected to end its current conditionality agreement with the IMF and was prepared to enter new business ventures.

At this time, Mexico was obliged to uphold its conditionality agreement with the IMF. The Mexican government had agreed not to raise the price for a barrel of crude oil above US$15. The IMF negotiated this arrangement on behalf of Mexico's creditors, with the US Treasury being the main financier (Flores-Quiroga 1998: 348; IMF EBS/90/10 1990: 42). Once the agreement expired, in 1990, Mexico expected to no longer enact similar conditionality agreements.

President Raúl Salinas de Gortari, who had taken office in 1988, pushed forward Mexico's relations with OPEC and attempted to erase the ill feelings of the past. He also attempted to diversify Mexico's economy by making it less dependent on the export of crude oil. To diversify, he attracted foreign investors into areas previously limited or closed off to foreign corporations (Chávez Ramírez 2006: 64). To do so, Salinas sold 50 percent of the state-owned agencies and began to relax federal regulations in various industries. His aim was to allow corporations to conduct business more easily and to make it clear to foreign investors that the government would not interfere in the operation of their businesses (see chapter 4). The area that attracted the most foreign investments, however, was the financial sector. Many investors chose to purchase high-interest Mexican treasury bonds rather than to establish businesses in Mexico. This was not the optimal method to expand employment opportunities in Mexico, but his plan did lead to the quick inflow of cash. The main investors were American citizens, who purchased 70 percent of the bonds (IMF SM/94/41 1994).

For the United States, the liberalization of the Mexican economy was seen as a positive change. Mexico's improved working relationship with OPEC was not met as a threat. Most likely this was a reflection of the US government's preparedness for future oil embargos. The US government had quickly raised the largest stockpile of petroleum in the world, and the SPR was sufficient to withstand any natural or politically created energy crisis. Oil-producing nations knew this. The SPR became a significant deterrent to oil import cutoffs by oil-exporting countries and a source of political power. The US Department of Energy projected that the SPR alone provided the nation with sufficient crude oil for about one year. In addition to the SPR, the US government had its own massive natural geological reserves. In 1989, the United States had the ninth-largest natural crude oil reserve in the world, consisting of 26.8 billion barrels of proven crude oil. Likewise, it could also rely on Mexico's crude oil reserves during a national crisis. Mexico held the eighth-largest proven reserve in the world, and in comparison to the United States had twice the amount of oil reserves (see table 5.1; table 5.2; appendix A). The United States estimated that Mexico's reserves could be as high as 54.1 billion barrels, while Mexico's estimates were more modest, at 46.25 billion barrels. As long as Mexico remained an ally and a productive oil-exporting nation, the US government had a dependable source to fill its consumption deficit. By 1990 Mexico had sold 56 percent of its daily oil exports to the United States, and the percentage continued to increase in the following years (table 4.1).

THE FIRST EXPERIMENT TO OPEN
PEMEX TO FOREIGN INVESTMENTS

In 1989 new discoveries in the Gulf of Mexico and along the southern states of Tampico and Campeche led Mexico to replenish its proven crude oil reserves. Twenty-four new deposits were discovered (Pemex, *Memoria de labores* 1989: 3; Pemex, *Memoria de labores* 1990: 3). These deposits helped Mexico slow down the depletion of its proven crude oil reserves, which had begun to decline from nearly 57 billion barrels in 1982 to around 48.6 billion in 1988 (table 5.2—US estimate).[5] To make Pemex more productive and to take advantage of the new discoveries, President Salinas asked the Mexican Congress to restructure the petroleum industry and allow private investments. He proposed that if Pemex improved its refining capacity of heavy crude oil into lighter blends, export profits would significantly increase, because a barrel of light oil was more expensive than heavy crude (Pemex, *Statistical Yearbook 1990*: 3, 36). To develop this infrastructure, Salinas argued, Pemex needed foreign financing.

The Mexican Congress was receptive to Salinas's proposal, but it was not prepared to revise the Mexican Constitution. This was a political battle that congressmen were not prepared to undertake at this time. Mexico was slowly emerging from its economic crisis caused by the US stock market crash of 1986, and entering a debate to privatize Pemex was not a good idea. Most Mexicans still disapproved of the revenue-raising privatization projects the Mexican Congress had recently enacted, and privatizing Pemex could potentially cause massive national protests. The president's main proposal, however, was accepted, and Congress allowed Salinas to restructure Pemex.[6]

To do so, President Salinas established Petróleos Mexicanos Internacional (PMI). Under the law, PMI was a subsidiary of Pemex, but legally it was a separate corporation that was not confined by the same regulations of the Mexican Constitution. PMI therefore was permitted to enter joint ventures with foreign oil corporations (Flores-Macias 2010). Although some of the president's critics argued that establishing PMI was possibly unconstitutional, Salinas was allowed to restructure Pemex (Manzo Yépez 1996). PMI was also mandated to manage the international commercialization of crude oil and petrochemicals. It was to directly work with importers and negotiate the price of crude oil, including coordinating policy with OPEC.

President Salinas began PMI's foreign business venture by entering partnerships with the Royal Dutch/Shell Oil Company and with Reposal. Both were petroleum corporations with long-standing relations with Mexico. The Royal Dutch/Shell Oil Company had been the largest investor in Mexican oil prior to the Mexican Revolution and had a history of complying with Mexican law (see chapter 2). It was a multinational corporation owned by Anglo-Dutch and US investors, and one of the six largest petroleum corporations in the world. Reposal, a Spanish firm, had a solid relationship with Mexico, as over the years it extended Pemex loans and had been Mexico's second-largest importer of petroleum products. The contracts between the corporations took effect in 1992. PMI and its partners were to split profits, risks, and investment capital in a 50-50 arrangement. Co-investments in the construction of high-capacity petroleum refineries were the first projects. Because the Mexican Constitution prohibited foreign investments in the petroleum industry, these jointly owned refineries could not be constructed on Mexican soil. It thus became necessary to build the plants in the United States and Spain. Once the refineries were established, crude oil would be transported from Mexico to the refineries and processed into lighter blends or petroleum products. Refined oil would remain stored in the refineries until transported to the purchasing countries.

PMI and the Royal Dutch/Shell Oil Company established their refinery in Houston, Texas, naming it the Shell Deer Park Refinery. Within a few years, it became Shell's second-largest refinery plant and the sixth-largest in the United States. In Spain, the PMI-Reposal refinery plant was established in 1993. This partnership, however, was not successful, and Mexico eventually came to own all of Reposal's assets. Reposal primarily served as a distributor for Mexican oil in Europe. By 2013, PMI would have established partnerships in twenty countries (PMI 2013) and entered other business ventures not related to oil.

For the US government, the establishment of the Shell Deer Park Refinery marked a new era in relations with Mexico. The Mexican government was not only showing signs that some congressional members favored opening the oil industry to foreign corporations but also allowing its oil to be stored on US soil. For the United States, this was an ideal situation, as it now had more oil at its disposal in case of a national emergency. The further liberalization of the oil industry, however, would have to wait at least two decades, since the majority of Mexicans, including those in the Mexican Congress, were against the privatization of the industry.

THE US-MEXICO OIL EXCHANGE AGREEMENT OF 1990

The 1989 crude oil discoveries in the Gulf of Mexico and along the southern states of Tampico came at an opportune time for the Mexican federal government, because these deposits indicated that Mexico had unseen potential. This was certainly good news for US consumers. The US government estimated that Mexico's reserves were much larger than Pemex's assessment, at least by 10 billion barrels. The Shell Deer Park Refinery established in Houston also demonstrated Mexico's goodwill toward the US government and proved that some Mexicans were in support of amending the Mexican Constitution and in favor of allowing foreign corporations to invest in Pemex. For the US government, both events were positive signs that Mexico's economy and energy policies were moving in the right direction. In January 1990, however, Mexico experienced a setback and informed its creditors that the federal government did not have the assets to meet its foreign debt installment payments. The federal government had a deficit of US$5.1 billion and needed financial assistance from the international community to pay its debts (IMF EBS/90/10 1990: 21). President Salinas asked the IMF to intervene on its behalf and assist Mexico to reschedule its loans. The federal government needed its short-term loans to be converted into medium- or long-term loans with payoff deadlines to be due from 1995 to 2006 (Pemex, *Informe Anual 2003*: Financial Appendix).

The IMF was able to reschedule Mexico's loans and negotiate for Mexico to receive new money. The US Treasury, however, required Mexico to accept several conditionality agreements before its agents signed the loan accord. First, Mexico was to immediately make a prepayment for the new money with a shipment of oil. The volume of oil was not disclosed to the public. Mexico also was to agree to extend the oil conditionality agreement currently in place until the US loans were liquidated (IMF EBS/90/10 1990: 14, 25). The price for a barrel of crude oil was to remain around US$15; if it deviated and reached $16.50, then Mexico's loan interest rates would rise. However, if the price of crude oil fell below $13, then Mexico would need to prepare a financial proposal proving that it would be able to immediately raise revenue to make its quarterly external payments. Therefore, before the Mexican government was allowed to renegotiate its old loans, it had to prepare a plan proving that if the federal government experienced another financial setback it could immediately institute new austerity policies to raise emergency funds (ibid., 25, 42, 56).

The IMF informed the US Treasury that in the case of "adverse exogenous developments," Mexico also had additional resources to raise emergency revenues (ibid., 33). The Mexican government owned several profit-making agencies, and if needed would sell these businesses, which included its communication, transportation, petrochemical, and fishery state agencies (ibid., 49, 51). The IMF assured US agents that Mexico was not opposed to privatizing more state agencies, since President Salinas had already taken the initiative to privatize the banking industry and sell Telemex, the state's telecommunication agency. The banks were to be privatized by no later than midsummer 1990.

President Salinas agreed to the US loan conditions, and his representatives signed the 1990 Oil Exchange Agreement. In March 1990, the first phase of the US oil exchange agreement was enacted and Mexico began its shipments of oil to the United States. A total of US$1.3 billion in crude oil was to be shipped and used for the US Strategic Petroleum Reserve (KAV 2553: March 23, 1990). The amount of oil exchanged was not disclosed to the public. The oil was to be delivered in shipments beginning in March 1990 and was to be valued according to the current conditionality oil price agreement of around US$15 a barrel. For the US government, the oil exchange agreement was profitable, as Mexico was obliged to sell its oil at a controlled price, which was lower than the global price (IMF EBS/90/10 1990: 25). For example, in 1991, US corporations and OPEC sold their oil at higher prices (see table 5.3). US corporate prices ranged from $18.46 to $21.54 per barrel of crude oil, and OPEC's average price was US$18.62 (OPEC 1999: 112, 115, 116). Following the oil exchange agreement, Mexico also significantly accelerated its daily exports of crude oil to the United States. In 1989 it exported 725.5 barrels a day, and by 1993 the amount had increased to 879.3 barrels (Pemex, *Statistical Yearbook 1994*: 32; table 4.1, table 5.4).

CAN PRIVATIZATION PROTECT MEXICO'S
PETROLEUM INDUSTRY?

In *Que Hacer con Pemex?* (1996), economist José Luis Manzo Yépez critically examines US-Mexican oil trade arrangements during the 1990s, arguing that the US government wielded undue influence on Mexican petroleum affairs. According to Manzo Yépez, this was carried out through political pressure imposed upon Mexico during the negotiation of financial agreements. Major loan transactions, Manzo Yépez states, require Pemex assets or its oil to be used as collateral insurance. Due to this structural problem, some members of the Mexican Congress have repeatedly suggested that to protect Pemex assets, privatizing Mexico's oil production is necessary.

Manzo Yépez concurs that Pemex assets must be reinvested in the industry and not used to meet foreign loan payment deficits, but he does not be-

TABLE 5.3. CRUDE OIL PRICES PER BARREL: MEXICO, OPEC, AND THE
UNITED STATES, 1990–2015 (IN DOLLARS)[1]

| Year | Mexico Export[2] | OPEC Basket Price[3] | United States[4] | | US Govt. Release SPR |
			WTI	ANS	
1990	19.09	22.26	24.53	21.92	
1991	14.58	18.62	21.54	18.46	Persian Gulf War
1992	14.88	18.44	20.58	18.16	
1993	13.20	16.33	18.43	16.37	
1994	13.88	15.53	17.20	15.73	
1995	15.70	16.86	18.43	17.18	Raise revenue
1996	18.94	20.29	22.12	20.53	Stabilize global prices
1997	16.46	18.68	20.61	18.93	
1998	10.17	12.28	14.42	12.50	
1999	15.62	17.49	19.34	17.77	
2000	24.79	27.60	30.38	28.40	Northeast winter storms
2001	18.61	23.12	25.98	23.34	
2002	21.52	24.36	26.18	—	Hurricane Lilly
2003	24.78	28.10	31.08	—	
2004	31.05	36.05	41.51	—	Hurricane Ivan
2005	42.71	50.64	56.64	—	Hurricane Katrina
2006	53.04	61.08	66.05	—	
2007	61.64	69.08	72.34	—	
2008	84.38	94.45	99.67	—	Hurricanes Gustav & Ike
2009	57.40	61.06	61.95	—	
2010	72.46	77.45	79.48	—	
2011	101.13	107.46	94.88	—	Libya Crisis
2012	101.81	109.45	94.05	—	Hurricane Isaac
2013	98.46	100.50	97.01	—	Mexican energy reform passes
2014	90.72	106.92	94.86	—	Secondary Laws approved
2015[5]	54.65	59.14	56.59	—	

[1]Prices are based on the international market and represent the average price of various crude oil types. Mexico subsidizes its domestic crude oil consumption by charging lower prices per barrel.

[2]Sources: Pemex, Statistical Yearbook 1999 (International Trade section) for 1990–1992; Statistical Yearbook 2003 for 1993–1999; Statistical Yearbook 2011 for 2000–2001; Statistical Yearbook 2013 for 2002–2012; Pemex 2014a for 2013–2014.

[3]Sources: OPEC Annual Statistical Bulletin 2001: 73 (Prices section) for 1990–2001; OPEC Annual Statistical Bulletin 2013: 87 for 2002–2012; OPEC 2014 (for 2014).

[4]Sources: WTI: EIA 2014a (for 1990–2010); EIA 2012 (for 2011–2012); EIA 2013e (2013); EIA 2014b (2014); ANS: OPEC 2001: 116.

[5]2015 data is for April 23, 2015 (Pemex 2015a; OPEC 2015). From January to May, the Pemex price per barrel of crude oil ranged from US$41.70 (January) to $57.27 (May 9) (Pemex 2015b).

lieve privatization should be Mexico's only alternative. He recommends that in the future, no matter what path the executive branch takes in its reform of the industry, Mexico should not permit US energy needs to dictate policy. To Manzo Yépez, the most effective approach in protecting Mexico's petroleum resources is to preserve the constitutional edicts that Mexico's hydrocarbons belong to the nation, and that the profits from the oil industry must be used solely in the interests of the people. He proposes that as long as the Constitution is not revised on these two points, the US government will respect Mexican law and not force the Mexican government to privatize Pemex.

To illustrate his argument, Manzo Yépez refers to the 1990 deliberations between the governments of the United States and Mexico, when the negotiations for the North American Free Trade Agreement (NAFTA) were under way (see chapter 4). During the talks, US diplomats proposed that Mexico first accept an oil-exchange agreement before proceeding with other points of discussion. Pemex was expected to export to the United States a minimum of 30 percent of its total crude oil production. If Mexico failed to meet this quota, then Pemex would be required to immediately cut its domestic sales and exports to other countries until the US quota was filled (Manzo Yépez 1996: 36). Manzo Yépez argues that Mexican diplomats were able to defeat this unreasonable demand by invoking Mexico's Constitution. He concludes that as long as Article 27 of the Constitution retains its basic principles, the Mexican government maintains the structural power to protect the sovereignty of its oil industry. As will be discussed in my forthcoming analysis of the 2013 energy reforms, antiprivatization activists advanced arguments similar to Manzo Yépez's and warned Mexicans that the reforms will weaken Mexico's ability to protect its sovereignty. They project that although the privatization of the oil industry did not take place, once the Mexican Constitution was revised and foreign investments in the oil fields permitted, Mexico's ability to retain control of the industry would be compromised.

In his internationally acclaimed book *Cómo la hacen de Pemex: La nueva guerra del petróleo* (2008), journalist Rafael Barajas Duran comically critiques past congressional attempts to privatize Pemex and offers an analogous assessment of US attempts to intervene in Mexico's oil industry. He goes beyond a constitutional assessment, however, and advances a postcolonial critique. Barajas Duran posits that privatizing the industry does not protect Pemex assets from foreign governments. It only turns over Pemex's future earnings to Mexican elites. According to Barajas Duran, Pemex is a profitable agency whose mandate is to use its earnings for the common good of the nation. Advocates who favor the agency's privatization have falsely claimed that Mexico will become an oil-importing nation unless it is converted into a for-profit corporation.[7] The intent of their argument is to place pressure on the Mexican government to privatize the industry, because the expansion of their fortunes depends upon forging business ventures with US investors. To Barajas Duran, there is no evidence to support the assertion that Pemex will become a more efficient agency under private control.

The main problem facing Pemex, according to Barajas Duran, is Mexico's oligarchy. Pemex is one of the last state agencies that is highly profitable, and which Mexican elites would like to own. They do not have sufficient capital, however, to purchase or manage the industry's daily costs. It is therefore necessary for Mexican investors to partner with foreign corporations before state ownership can be terminated. Barajas Duran concludes that the movement to privatize Pemex parallels past historical moments, similar to the days of the Porfirian period, when the ruling class did not care about the people's social welfare. He concludes that ultimately the privatization of Pemex is not a devious plan constructed by the US government, but rather a domestic issue reflecting the greed of Mexico's upper class.

In *Pemex: La reforma petrolera* (2005), David Shields identifies mismanagement, rather than capitalist greed, to be Pemex's main problem. Shields argues that historically, Mexico's executive branch has mismanaged the petroleum industry by excessively taxing Pemex. Pemex's ability to reinvest its profits in constructing new refineries, modernizing technology, and expanding its exploratory capacity is highly diminished by the government's tax regime. Shields advocates for Pemex to be transformed into an independent corporation whose assets cannot be used to fund the federal budget. Regardless of whether the model for independence is privatization or a structure where the agency remains under state control, it is critical that Pemex be allowed to reinvest its profits. Shields warns that if Pemex continues to be excessively taxed, and if its production level continues to be determined by oil-exchange agreements, then Mexico's crude oil will be depleted. In sum, while Shields is inconclusive on the restructuring plan Pemex should undergo, he is convinced that treating Mexico's hydrocarbons as a sacred cow to be used solely in the interest of the people is an outdated premise that will only lead to the agency's financial ruin.

Whereas Manzo Yépez, Barajas Duran, and Shields offer different scenarios of the ethics and politics of the privatization of Pemex, they concur on one central point: since the 1980s, Mexico has been overproducing crude oil and the people of the United States have been the main beneficiaries. They argue that the US government has unfairly pressured the sitting Mexican presidents to accept oil-exchange agreements that have harmed Mexico's oil industry. Mexico has not been given alternate arrangements to resolve its financial problems, other than to mortgage its oil deposits. Although I concur that the outcomes of the oil exchange accords were negotiated during periods of economic stress and between partners of unequal power, I propose that the Mexican government ultimately accepted the conditions in efforts to remain in good standing within the global economy and to resume its credit eligibility.

THE OIL AGREEMENT OF 1995: MEXICO'S ENERGY INDEPENDENCE BECOMES IMPOSSIBLE

In the mid-1990s, the possibility of ending the 1990 conditionality agreement setting Mexico's oil prices became a virtual impossibility as Mexico experienced the most severe economic depression in its history. As discussed in chapter 4, the primary cause of the economic crash was the insolvency of Mexico's recently privatized banks. In December 1995, the banks collapsed (IMF CR/4/418 2004; IMF EBD/94/200 1994; Ros and Bouillon 2001: 727). The federal government and the privatized banks had recklessly sold long-term treasury bonds at high interest rates ranging from 14 to 37 percent (IMF SM/94/41 1994: 46, 59; INEGI 1996: 163).[8] This project was part of Salinas's privatization movement to boost the Mexican economy by enticing foreigners and domestic elites to purchase treasury bonds and notes. Salinas, however, had not envisioned that investors would cash in their securities at the first sign of trouble.

When news spread that neither the government nor the banks had sufficient assets to reimburse the treasury bonds and notes, people panicked and began withdrawing their savings, leading to the collapse of the banks and the economy. Exacerbating matters, the federal government did not have the funds to meet its US\$28 billion external debt payment. With the economy in ruins, thousands of businesses went bankrupt and many people lost their jobs.

The US government immediately intervened to stabilize the Mexican economy, for if it did not, it expected a massive surge in undocumented immigration and serious financial losses for US banks, because the majority of the Mexican bonds and notes were owned by US banks and American investors (GAO 1996: 6, 11, 64). President Clinton prepared a bailout package totaling US\$40 billion. Since the Mexican government was bankrupt and its only major source of collateral was Pemex's future earnings, Clinton prepared a new oil agreement. This plan replaced other oil-exchange agreements made with the US government. Clinton's goal was to stabilize the Mexican economy, but also to make sure that Mexico had the assets to repay investors and meet its new and past quarterly foreign debt payments. Under the agreement, representatives of the US Treasury were appointed to closely monitor

Mexico's crude oil industry and set mechanisms in place to maintain Pemex's productivity. Under the Oil Proceeds Agreement of 1995, Mexico agreed not to raise its prices above 15 percent of the current amount set by the United States, nor to lower its prices below 25 percent (GAO 1996: 2; KAV 4268; KAV 4269). Mexico was required to retain this agreement for ten years, or until the 1995 debt was paid. Mexico also agreed to exchange an undisclosed amount of crude oil based on 1990 prices (Pemex, *Informe Anual 1997*: 49). Furthermore, Mexico was required not only to be under the surveillance of the US Treasury but also to work with the IMF in fulfilling other conditionality agreements placed by creditors.

As Mexico's oil industry was to be closely monitored by the US Treasury, the Mexican government initiated a plan to quickly pay off the 1995 debt. The payment of other loans did not have to be accelerated, since conditionality agreements monitoring the oil industry had not been required. To raise revenue for its external payments, the federal government accelerated the amount of crude oil for export. In 1994, Mexico exported 1,307.4 million barrels a day; by 1997 the amount had increased to 1,720.7 million barrels (table 5.4). During this time, Mexico's main importers continued to be US corporations as well as the US government, who together purchased from 73 to 79 percent of Mexico's total crude oil exports. The cost of Mexico's oil was significantly lower in comparison to global market prices. For example, from 1996 to 1997, US corporations sold a barrel of crude oil from US$22.12 to $18.93, and OPEC from $20.29 to $18.68 (table 5.4, table 5.3), whereas Mexico's prices ranged from $18.94 to $16.46. Although US petroleum corporations reaped the benefits of the oil-exchange agreement by having an abundant source of inexpensive oil, there is no question that the recovery plan salvaged Mexico's economy. Without the US government's aid, Mexico's crisis would have devastated the nation and caused the poverty level to rise above the 69 percent level already in place.

In 1997 the Mexican government raised sufficient funds to pay off the Oil Proceeds Agreement of 1995. Before the Mexican government was allowed to liquidate this loan, it had to consent to several conditions. President Zedillo agreed to increase the federal government's long-term interest rates for loans issued in 1983, 1987, and 1990 (Pemex, *Informe Anual 1998*: 62; Pemex, *Informe Anual 2003*; Pemex, *Statistical Yearbook 1996*: 57). The US Treasury also required Mexico to exchange crude oil instead of cash as part of the prepayment (Pemex, *Informe Anual 1997*: 49).[9] The amount was not publicly disclosed. Mexico also paid the US Treasury US$6 billion in cash, which it borrowed from the Deutsche Bank (ibid., 57). Paying the loan early and switching creditors for part of the loan was beneficial to Mexico, since it was no longer obligated to have the US government monitor its oil sales. The

TABLE 5.4. PEMEX CRUDE OIL PRODUCTION AND EXPORTS,
1993–2015 (THOUSANDS, DAILY BARRELS)

Year	Production Total	Export Total	Export to US	Percent of Total Export to US	Average Price (in US Dollars)
1993	2,673.0	1,337.1	879.3	66	$13.20
1994	2,685.0	1,307.4	960.8	73	$13.88
1995	2,617.0	1,305.5	1,037.1	79	$15.70
1996	2,858.3	1,543.8	1,209.6	78	$18.94
1997	3,022.2	1,720.7	1,334.9	78	$16.46
1998	3,070.5	1,741.2	1,341.5	77	$10.17
1999	2,906.0	1,553.5	1,171.2	75	$15.62
2000	3,0120	1,603.7	1,203.4	75	$24.79
2001	3,127.0	1,755.7	1,321.7	75	$18.61
2002	3,177.1	1,705.1	1,338.6	79	$21.52
2003	3,370.9	1,843.9	1,437.5	78	$24.78
2004	3,382.9	1,870.3	1,482.0	79	$31.05
2005	3,333.3	1,817.1	1,424.7	78	$42.71
2006	3,255.6	1,792.7	1,441.9	80	$53.04
2007	3,075.7	1,686.2	1,351.5	80	$61.64
2008	2,791.6	1,403.4	1,142.9	81	$84.38
2009	2,601.5	1,222.1	1,049.0	86	$57.40
2010	2,575.9	1,360.5	1,139.5	84	$72.46
2011	2,552.6	1,337.8	1,094.9	82	$101.13
2012	2,547.9	1,255.6	957.1	76	$101.81
2013	2,522.1	1,189.0	856.9	72	$98.46
2014	2,520.0	1,142.0	828.0[1]	73	$90.72
2015	—	1,282.0[2]	—	—	—

Sources: Pemex, *Statistical Yearbook 2001* (sections Exploration and Production, International Trade) for 1993–2001; *Statistical Yearbook 2011* for 2000–2010; *Statistical Yearbook 2012* for 2011–2012; *Statistical Yearbook 2013/14* for 2013–2014; Pemex 2014a for 2013; Pemex 2015b for 2014–2015; see also EIA 2011 (US exports 2002–2009).

[1]For 2014, EIA reports 981,000 barrels of daily exports, which is more than Pemex reports (EIA 2014c).

[2]2015 data is for January and February.

Deutsche Bank had not placed a comparable oil conditionality agreement. Although Mexico closed this chapter of its financial history, its external debt fell by only US$5.2 billion, since a large percentage of the loan was switched to another creditor rather than being paid off. In 1997, Mexico's external debt amounted to $160.2 billion (Girón González 2001: 255; IMF CR/11/250 2011; World Bank 2013).

Mexico's early-payment plan was costly and had adverse consequences on Pemex. As part of the US loan negotiation conditions, Mexico agreed to allow the US Securities and Exchange Commission (SEC) to assess Mexico's proven crude oil reserves. The US government was willing to allow Mexico to nullify its 1995 agreement, yet it was not willing to completely stop monitoring Mexico's most valuable resource, which was crude oil. The SEC became empowered to evaluate Pemex crude oil estimates and to either certify or downgrade current calculations. For US investors, it was important to have the SEC intervene in Mexico's oil business. The SEC, on behalf of foreign lenders, used Mexico's crude oil reserves as an indicator of the Mexican government's ability to repay its loans (Shields 2003; Society of Petroleum Engineers 2005). The amount of reserves was used to project Mexico's potential annual oil sales and profits. At this point in time, Mexico was required to temporarily adopt the SEC's evaluation standards to measure reserves, for if it did not, the commissioners could decide that Mexico's estimates were inconclusive, which could damage or nullify financial negotiations with potential foreign creditors. Thus, in 1997 Pemex entered talks with the SEC to develop evaluation standards. Mexican engineers disagreed with the SEC standards, yet they were forced to accept the new reevaluation measures. Within two years, Mexico's crude oil estimates were downgraded by the SEC, culminating in a 41 percent decrease (Krauss 1997; Pemex, *Statistical Yearbook 1999*; table 5.2; appendix A). The next president, Vicente Fox, did not easily accept this new form of surveillance and questioned the accuracy of the SEC's measurements, since the downgrade was largely based upon Mexico's financial inability to extract deep ocean deposits rather than on its geological findings.

REFORMING ARTICLE 27 OF THE MEXICAN CONSTITUTION AND THE MEXICAN ECONOMY BEFORE PRESIDENT FOX'S ADMINISTRATION

The political ramifications of Mexico's economic crash in 1994 included the rise of a nationalist movement to prevent the privatization of Pemex. By the end of Zedillo's presidency, most state agencies had been privatized and average people did not feel as though they had benefited (see chapter 4). Although the recovery plan helped to stabilize Mexico by infusing American dollars into the economy, inflation continued to increase (INEGI 2000b: 49). The poverty level also remained above 60 percent (Consejo Nacional de Población 2009: 25). Popular sentiment was that the privatization of the state agencies had been a financial disaster; therefore, for the welfare of the nation it was important to ensure that Pemex remain a state non-profit corporation (Flores-Macias 2010; López Obrador 2010). When state agencies were sold, the cost of living increased radically. Food, particularly staples such as tortillas, became more expensive. Transportation and communication costs also skyrocketed. Telephone, airfare, and train fares more than doubled, and in many parts of the nation highway tolls were charged. People also complained that it was difficult to qualify for credit, and even when it was available, high interest rates prevented most Mexicans from using their credit cards.

As the last recourse to the nation's problems, the Mexican Congress had reluctantly supported President Zedillo's privatization practices, since raising revenue via privatization had been part of the conditionality agreements placed upon Mexico. In helping Mexico raise revenue and reduce government expenditures, the IMF advised privatizing the last state-owned enterprises. In particular, it was important to sell the businesses that were costly to maintain, such as federal highways, ports, Pemex, and the electrical industry (GAO 1996: 22; Girón González 2001: 245, 249, 271). By the end of President Zedillo's term, only eighty-one fully owned state agencies were left (Chávez Ramírez 1996: 151, 153). To rapidly infuse capital into Mexico, IMF staff also advised the president to permit foreign investors to own a higher percentage of the banking industry, in order to restore the flow of credit (IMF BUFF/ED/94/27 1994). In 1995, following the enactment of the bailout agreement, foreign investors were allowed to own 51 percent of a Mexican bank (up from 30 percent in 1991), and four years later all limits were removed, permitting full ownership by foreign firms (Bancomer 2004: 18).

In response to IMF advice, President Zedillo introduced reforms in the energy field. He asked the Mexican Congress to allow private investments in Pemex and the electrical industry. Congressional representatives, however, mainly supported reforms in the electrical field and minor changes within Pemex (Fundación Colossio 2008). On May 11, 1995, Article 27 of the Mexican Constitution was revised, allowing the government to contract private firms to transport, store, and distribute natural gas (Flores-Macias 2010). Previously, only Pemex was authorized to conduct these services. Most other reforms were enacted in the electrical industry. Zedillo's administration began to partner with private investors to fund electrical infrastructure improvements under the "piedriega program" (i.e., construction projects, electrical grid upgrades). To attract investors, the federal government assumed the entire risk if a project failed or was unprofitable. However, if the project succeeded, the partners shared the profits. IMF staff informed the president that this was not the best approach to financing infrastructure improvements, since the private sector did not assume any risk and was not held responsible for completing a project if something went wrong. This type of financing was risky and could sink the government in heavy debt (IMF CR/4/418 2004). IMF staff instead advised the president to allow domestic and foreign corporations to invest in deep ocean oil drilling projects where deposits were known to exist and could generate enormous profits. Currently, Pemex did not have the finances or the technology to drill in some of its deep ocean reserves.

PRESIDENT FOX AND THE CLAIMS
OF DEPLETING OIL RESERVES

Under the administration of Vicente Fox, who entered the presidency in 2000, Mexico did not enact any initiative to privatize Pemex. The president attempted several times, however, to negotiate an oil-exchange agreement that was favorable to the United States, if in turn the US Congress renewed the bracero program (see chapter 4). Beginning in 2002, Fox offered to accelerate Mexico's production level and increase oil exports to the United States. Mexican activists accused Fox of placing Mexico in the path of turning into a crude oil–importing country, as Pemex's proven reserves in the previous ten years had fallen from eighth in the world (1996) to fourteenth, yet the government had not reduced production (table 5.1; Flores-Macias 2010; Shields 2005). Mexico remained the sixth-largest exporter of oil in the world, regardless of its declining reserves. If this pattern persisted, economists projected that Mexico's reserves would be completely depleted.

Those who supported President Fox's export plan disagreed with the depletion doomsday prediction, as they believed Mexico's proven reserves were much larger than accounted for (see Díaz Serrano 1992; Leos 2006). From 2002 to 2004, the US Securities and Exchange Commission had once again pressured Mexico to reduce its proven crude oil reserve estimate after US geologists proposed that approximately half of Mexico's proven reserves were only probable reserves (EIA 2013c; Shields 2003). Mexico estimated its proven reserves were around 21.89 billion barrels in the winter of 2002, while the SEC disputed this claim and put the number at 12.6 billion barrels (appendix A; table 5.2). Proven reserves could be easily extracted from shallow deposits, while probable deposits required a more expensive drilling process. Although Mexico's crude oil reserves were extensive, the SEC instructed Pemex to downgrade its proven reserve estimate and reclassify a large percentage of it into probable reserves, because Mexico did not have the technology or finances to recover its deep ocean oil deposits. This determination affected Mexico's world standing and its ability to borrow money based on the amount of its proven crude oil reserves.

Although Pemex followed the commission's mandate to recalculate Mexico's proven oil reserves, President Fox disputed the US estimate and refused to reduce Mexico's production level. He also determined that this was not the time to reduce production, since the price of oil was rising rapidly (Leos 2006; Shields 2005). When Fox began his presidency, in 2000, the price of a barrel of crude oil was US$24.79; at the end of his administration, in 2006, it had increased to $53.04 (table 5.3).

Thus even though the US government refused to negotiate an oil-exchange agreement during Fox's presidency and Mexico's international clout fell because its export sales were projected to decrease, Mexico's oil sales did well. Its annual export profits more than tripled due to the high price of crude oil. Mexico was able to export less oil, yet charge higher prices for what it sold. For example, in 1998 Mexico's crude oil export revenue from the United States was US$5.1 billion for 488.5 million barrels of crude oil; in 2006 revenue climbed to $28 billion for nearly the same amount of exported oil (525.7 barrels of crude oil) (table 5.5).

With Mexico making substantial profits from its crude oil exports, the IMF advised President Fox to conserve the funds. Mexico's external debt had reached US$169 billion, and if the economy experienced a new setback, it would be difficult to pay US creditors (IMF CR/11/250 2011). Fox refused to listen and instead increased federal spending. He lobbied the Mexican Congress to pass LAW 112, which allowed the federal government to increase the public budget when Pemex's profits exceeded the projected annual target (IMF CR/5/427 2005; IMF CR/6/352 2006). With the added cash flow from the oil exports, Fox relaxed the federal austerity policies imposed upon the people by previous administrations. The president extended loans and grants for infrastructure projects to municipalities and state governments (IMF CR/4/418 2004). Under Fox's administration, Mexico's poverty level declined; it remained high but fell considerably, reaching 43 percent by 2006 (World Bank 2012). Although Mexicans enjoyed a temporary relief from institutionalized austerity policies, the president's tax policies put Mexico deeply in debt. At the same time that he passed policies to benefit the masses, Fox also forgave the taxes of many corporations if their profits were below expectations (López Obrador 2010). This reduced the government's tax revenues and caused the government deficit to grow.

TABLE 5.5. MEXICO'S ANNUAL CRUDE OIL EXPORT SALES TO THE
UNITED STATES, SELECTED YEARS

| Year | Total Exports | US Exports Barrels in Millions | | US Sales (in US Billions) | Historical Period |
		No.	Percent of Total		
1995	472.3	375.5	80	5.9	Recovery plan agreement
1997	627.3	489.0	78	8.2	Recovery agreement ends
1998	633.3	488.5	77	5.1	
2000	587.0	440.4	75	11.1	
2001	641.0	482.4	75	9.0	9/11 attack on US
2002	622.4	489.2	79	10.6	
2005	663.3	521.1	79	22.3	US House of Rep. immigration debate
2006	654.3	525.7	80	28.0	US Senate immigration debate/ end of Pres. Fox's term
2008	514.0	417.3	81	35.4	Privatization debate
2011	488.3	400.0	82	40.3	

Sources: INEGI 2012a: cuadro 21.11; INEGI 2013a: cuadro 22.11.

Before President Fox left office, the IMF warned that Mexico needed a quick infusion of cash if its economy was to continue running smoothly. It was advisable to raise revenue by allowing private investments in Pemex (IMF CR/4/418 2004; IMF CR/4/419 2004; IMF CR/6/352 2006; IMF CR/11/250 2011). Fox listened and set plans in motion, but he left this issue to the next president (Leos 2006; Shields 2005).

FELIPE CALDERÓN'S INITIATIVE TO PRIVATIZE PEMEX

In 2006 Felipe Calderón, Fox's secretary of energy, became president following a highly contested election. Calderón, a Harvard-trained economist, ran against Andrés López Obrador, who was the head of government of the Mexican Federal District (Mexico City) and a populist liberal politician. López Obrador ran on a social justice platform, promising to reform the government's neoliberal policies by ending corporate welfare, which he believed led to rising poverty rates. Although López Obrador lost the election, Calderón's presidency was discredited and lost legitimacy. His moral authority over the nation worsened after he unsuccessfully attempted to privatize Pemex.

In 2008, President Calderón began proceedings to privatize Pemex, despite widespread public opinion against this plan (Barajas Duran 2008). He claimed that privatization would infuse Pemex with needed capital and increase its productivity. According to the president, Mexico's probable reserves needed to be explored and made productive. For years Mexico's proven reserves had been declining, and it was necessary to invest in developing Mexico's probable and possible reserves (see appendix B). Mexico's crude oil exports supposedly could be doubled if his plan was implemented. The US government, the IMF, and the Foreign Relations Council supported Calderón's plan. Privatization would proceed by: 1) selling shares in Pemex to domestic and foreign investors and 2) selling Pemex assets, such as refineries, ducts, oil platforms, and machinery.

Opposition to privatizing Pemex centered on one main argument: if Mexico privatized the oil industry, then it would lose its most valuable commodity and main tax revenue. Historically, the government had relied on Pemex taxes to finance around 40 percent and sometimes up to 60 percent of the federal budget (Fundación Colossio 2008: 92; INEGI 2009a: 208; Manzo Yépez 1996: 51). The critics agreed that in the short term, the influx of capital would benefit Mexico, but they projected that in the long term, this tax base was irreplaceable.

A group of academics calling themselves the Intellectual Committee in Defense of Petroleum opposed Calderón's proposal and submitted their critique to Congress (Cárdenas Garcia 2009). They projected that two of the proposal's provisions would cause Mexico to lose control of Pemex to foreign corporations. Article 6 of Calderón's proposal was alleged to potentially transform Pemex into a foreign-owned industry, with most profits leaving Mexico, because the number of shares sold to foreign investors was unrestricted. Article 49 was also denounced for transferring control of Pemex to foreign interests, since disputes over investments would be arbitrated in foreign courts.

In the meantime, López Obrador organized supporters, and throughout Mexico, protests and referendums to stop the privatization of Pemex succeeded (Flores-Macias 2010). On August 28, 2008, Mexico's Congress rejected privatization based on the finding that shared public-private projects in the energy field often failed and were less efficient than those solely controlled by the government (Fundación Colossio 2008; IMF CR/6/352 2006). The piedriega program that President Zedillo had instituted proved to be a failure (Shields 2005: 4). As the IMF had forewarned Mexico, the federal government assumed the entire risk in the piedriega infrastructure projects, and when they failed, it was difficult to hold private investors accountable, since they could walk away from a project without incurring financial losses. By 2004, many projects had failed, and 43 percent of Pemex debt was due to piedriega projects (IMF CR/4/418 2004).

Although Congress chose to not privatize Pemex at this time, it did adopt part of Calderón's reforms. Shares in Pemex were sold and called petrobonds. The shares, however, could be sold only to Mexican citizens (Pemex, *Informe Anual 2008*; Perez-Rocha and Schwartz Grego 2008). Congress also made it clear that the petrobonds did not give shareholders any ownership in Pemex or in Mexico's oil. All hydrocarbons, including oil, and the property from which they were extracted remained the sole property of the nation. Calderón's main victory was a moderate revision of Article 27. Congress approved incentive-based contracts, which removed Pemex's exclusive right to explore for oil in Mexico (Martin and Rodriguez 2012). Foreign and domestic companies could be contracted to explore and drill oil wells for a variable fee.

Critics protested, charging that the variable fee was similar to investing in Pemex since an incentive-based contract gave the partnering corporation the right to claim a percentage of the profit when oil was discovered. The privatization debate did not end. Opponents of the president's proposal raised constitutional issues and took their case to the Mexican Supreme Court, while supporters of privatization promised to radically reform the Mexican Constitution and end the nation's ownership over Mexico's hydrocarbons. In December 2010 the Supreme Court rendered its decision, upholding the congressional decree. Critics bemoaned the loss and perceived this to be the first phase in privatizing Pemex (Guthrie 2010).

During President Calderón's administration, Mexico's annual crude oil export sales continued to increase in value as the price of oil nearly doubled (see tables 5.4 and 5.5). This was a positive trend for Mexico and a very costly outcome for the United States. Mexico was selling less oil than in previous years but making double the profit. In 2011, a year before President Calderón left office, Mexico's total annual export volume had fallen to 488.334 million barrels of crude oil, with US importers purchasing 82 percent of the oil for a total of US$40.3 billion (table 5.5). US costs for importing oil had increased by 65 percent in the previous five years. In 2006 it perhaps would have been cost-efficient for President Bush to have accepted President Fox's oil agreement in exchange for renewing the bracero program (see US Senate Foreign Relations Committee 2006).

Mexico's profitable earnings were due to the high cost of a barrel of crude oil, which had peaked at US$101.13 in 2011, in comparison to $53.04 when Calderón entered the presidency (table 5.3). Although Mexico's revenues were outstanding, its production level significantly declined during the Calderón administration When Calderón left office, in 2012, Mexico cut back production from its 2006 level and fell from the sixth-largest crude oil producer to the tenth (table 5.1). Its exports to the United States also fell to 76 percent of total exports. This caused global speculation as to why Mexico's oil production and export practices had changed (Rodriguez and Roeder 2012). Was Mexico choosing to conserve its crude oil, were its reserves depleted, or did Mexico not have the technology to extract its deep ocean deposits?

MEXICO'S PRESIDENTIAL ELECTION OF 2012: TWO PATHWAYS TO IMPROVE A DEPRESSED ECONOMY

Toward the end of President Calderón's presidency, Mexico's economy had certainly recovered from the troubles it had experienced in the past. In 2012, Mexico's GDP growth of 3.4 percent was impressive in comparison to the −6.2 percent deficit growth of 1995, when the economy crashed following President Salinas's ineffective privatization and deregulation growth policies (IMF CR/12/317 2012: 6; IMF CR/13/334 2013: 28; Tello 2009: 257). In 2012, Mexicans could also boast that their country was the home of Carlos Slim, the wealthiest man in the world, who during Salinas's term became ultra rich after purchasing the state-owned telephone company, Teléfonos de México (Kroll 2012; Martínez 2011).[10] Although Mexico's economy had gradually improved over the years, the poverty rate was still a problem and had worsened during President Calderón's administration. According to the IMF and the World Bank, Mexico's poverty rate had fallen to 43 percent in 2006, but by 2011 it had risen once again, to 51.3 percent (IMF CR/11/250 2011; World Bank 2012).[11] Economist Carlos Tello, a former director of Banco de México (the equivalent of the US Federal Reserve), described Mexico as a nation of extreme poverty, where 10 percent of the poorest households survive on 1 percent of the nation's assets, while 10 percent of the wealthiest households own 40 percent of the nation's wealth (Kahn 2013; Tello 2009: 253).

In the summer of 2012, Mexican citizens were given the opportunity to elect a new president. Their candidates held very different visions for promoting Mexico's prosperity and addressing the nation's social ills. The frontrunners were Enrique Peña Nieto and Andrés López Obrador. Peña Nieto was the presidential candidate for the Partido Revolucionario Institucional (PRI). He ran on what appeared to be a social class–neutral campaign, promising to be a fiscal conservative, but also affirming his commitment to restructuring the economy by increasing Mexico's potential to raise revenue. To this end, his main priorities were to attract foreign investments to Mexico, increase taxes on corporations and the highest income earners, and restructure Pemex by converting it into a profit-making state agency. Essentially, his populist agenda promised to help all economic sectors, while at the same time keeping the private sector accountable to the state.

Peña Nieto's campaign platform differed radically from the platform of López Obrador, the candidate for the Partido de la Revolución Democrática (PRD). López Obrador had chosen to once again run for office. His platform promised to reduce poverty, improve the schools, fight government corruption, and restructure Pemex, but without privatizing the industry. When voters cast their ballots on July 1, 2012, the results were close, but not nearly as close as six years earlier, when López Obrador had run against Calderón with less than a 1 percent margin of difference (Calderón won 35.86 percent of the vote and López Obrador 35.31 percent; see CNN 2012). In 2012, López Obrador obtained 32 percent of the vote, the Partido Acción Nacional candidate Josefina Vásquez Mota 25 percent, and Peña Nieto 38 percent (Padgett 2012). Because none of the candidates obtained a majority of the vote, López Obrador demanded a recount. He also raised corruption charges against Peña Nieto, accusing his operatives of illegally purchasing millions of votes by issuing impoverished citizens gift cards to purchase food and merchandise.

López Obrador charged that millions of Monex gift cards were given to voters nationwide who were willing to sell their vote in return for money to purchase food and other basic necessities. In states where the elections were close, such as in Puebla, Tabasco, and Veracruz, gift cards of higher value were given, with Bancomer and Sandanter cards containing two thousand to four thousand pesos. Likewise, in the border states of Sonora, Chihuahua, and Tamaulipas, where elections were projected to be close, telephone cards and Soriana food cards were exchanged for votes (Padilla 2012: 4). The Electoral Tribunal of the Federal Judiciary investigated the charges and concluded that in a few localities, fraud may have taken place, but it had not occurred on a grand enough scale to tip the election results (Jimenez 2012: 4; Mercado 2012: 3). On August 30, 2012, the electoral tribunal certified Enrique Peña Nieto's presidential win, rejecting the charges against him and his PRI Party.

MEXICO'S ECONOMIC OUTLOOK: THE OBSTACLES PRESIDENT PEÑA NIETO MUST ADDRESS

The Mexican Census of 2010 indeed supports Carlos Tello's analysis that Mexico is a country of extreme poverty. In 2013, 59.1 percent of Mexicans were described as living in poverty (*El Economista* 2013; INEGI 2013b).[12] Mexico's census distinguishes the poor as belonging to the upper working class or the lower working class. Those in the lower working class live in extreme poverty and are characterized as having insufficient food and inadequate housing and clothing. This sector is estimated to constitute 35 percent of the Mexican population (*Forbes Mexico* 2014). Those in the upper working class also live in poverty, but they have sufficient income to purchase food and pay for housing.

In 2013 the Instituto Nacional de Estadísticas y Geografía concluded its analysis of Mexico's economic gains, basing its assessment on the Mexican Census of 2010. INEGI offered an optimistic but mixed assessment of the nation's social-class mobility. In the previous ten years, Mexico's middle class had increased to 39.16 percent, growing by 11.4 percent of the total population (Zuñiga 2013: 1). Therefore a large percentage of the Mexican population lived comfortably and had enough earnings to purchase commodities beyond their bare necessities. The question that remained unanswered is, why did this sector increase? Did a significant number of the upper working class enter this sector, was there a significant drop in the upper class, or did the middle class expand because young adults retained their parents' status when living on their own?[13] Most likely the latter is what occurred, since the percentage of Mexico's upper class remained unchanged and data on Mexico's poor does not support the assessment that a large percentage of the working class moved out of poverty. The numerical size of Mexico's poor actually had increased in the previous ten years (INEGI 2013b; Ortiz Santillán 2013).[14]

The International Monetary Fund and the World Bank offered an assessment similar to INEGI's. Mexico had made great advancements, yet it still needs to improve various areas of its economy, specifically in the schooling of its population, since education is the gateway to social mobility and the main vehicle to improve a nation's technological level, which ultimately translates into economic gains (Corbacho and Schwartz 2002; IMF CR/10/70 2010; IMF CR/11/250 2011: 31; World Bank 2005, 2012).[15] President Fox had greatly improved Mexico's educational spending, ending the limits Zedillo and Salinas had placed in this area, yet Mexico still lags behind most other Latin American nations. The World Bank found that Mexico has the lowest gross investment in primary education among Latin American and Caribbean countries (World Bank 2013).

Mexican academics concur with the World Bank's assessment and argue that most educational advancements have been in the financing of basic education, which consists of primary and secondary schooling (the equivalent to US elementary to junior-high education) (López Obrador 2010). Approximately 66.6 percent of the educational budget is used to finance basic education (INEGI 2009a: 80; INEGI 2012a: cuadro 4.35).[16] The problem with this scenario is that mid-level education, which is called *preparatoria* (general or technical education) and is similar to a high school education in the United States, is seriously underfunded. In 2010, only 5.8 percent of the educational budget was used to fund this level though more than one-third of youth are of age to attend *preparatoria* schools.

Many critics of the federal government's educational policy argue that publicly funded education in Mexico is not intended to serve youth past age fifteen. As evidence, they identify the passage of the educational reforms of 1992, which revised Article 3 of the Mexican Constitution, as a clear indication that mid- to higher-level education was no longer a governmental priority. In 1992 the federal government redefined its commitment to public education and guaranteed funding only for basic education (López Obrador 2010: 26, 100). Following the reform, school attendance dropped 36 percent from 1992 to 2009 for students ages fifteen to seventeen. The World Bank estimates that the drop in school attendance for the extremely poor of the same age group was more severe, as by 2004 less than 55 percent of this group attended secondary school (World Bank 2005: 121). Educational opportunities had not improved by 2010. Over 60 percent of youths ages fifteen to nineteen were not attending school (IMF CR/10/70 2010; INEGI 2012a: cuadros 4.9, 4.26).

The problem with this educational scenario is that most youths cannot enter formal employment because the government does not produce sufficient jobs to integrate them. If youths need to work, then they must be prepared to enter service occupations. In Mexican public discourse the term *nini* has become a popular idiom to describe Mexico's current generational employment problem. *Ninis* are youths who are unable to find work or attend school, meaning "ni escuela, ni trabajo." Mexico's main occupational growth continues to be in the service sector. In 2012 the percentage of manufacturing jobs remained constant at around 23.1 percent, nearly the same level as in 1970 (24.4 percent) (IMF CR/13/334 2013: 28; INEGI 2000b: 115; INEGI 2009c: 14; INEGI 2012b: 16; Moreno-Brid and Ros 2009: 184). Making matters worse, of those employed in manufacturing, the majority are not salaried and most workers are on call and paid by the hour (INEGI 2012a: cuadro 9.13). The World Bank estimates that in 2011, Mexico had the twelfth-largest economy in the world, yet in Latin America it also ranked as the nation with the lowest job production level for a middle-income country (World Bank 2005: 114, 2012).

With respect to the informal labor market, the federal government estimates that in the last thirty years, employment in this sector has continued to rise. Informal employment is not just a problem for the working poor who do not earn a living wage; it is also a serious problem for the federal government, since these types of laborers cannot be taxed (see chapter 4; IMF CR/11/250 2011: 31). In 2010, approximately 29.3 percent of the working-age population was self-employed in low- to no-income informal occupations (INEGI 2012b: 17). Since then, INEGI surveys indicate that informal work is on the rise, with six out of ten workers stuck in this occupational sector (Dinero 2013).

Whether President Enrique Peña Nieto, who took office on December 1, 2012, can reverse these educational and economic patterns by instituting structural reforms is a question that only time can answer. He faces a difficult situation, as the IMF reported that in the first months of his administration, Mexico experienced a severe and unexpected deceleration of the economy, with the GDP falling from 3.4 percent growth in 2012 to 0.3 percent growth in June 2013 (IMF CR/13/334 2013: 2). The economic stagflation was attributed to the contraction of the US economy.

PROJECTIONS: MEXICO'S OIL INDUSTRY
WILL STABILIZE THE ECONOMY

President Peña Nieto was to unveil his economic plan for the nation in the summer of 2013. Reforms in Mexico's oil energy sector were expected to be the centerpiece of his legislation, since the effects of the US economy's contraction were to be counterbalanced by income generated by oil exports. Pro-privatization advocates waited to hear that the president planned to privatize the oil industry as the main measure to address Mexico's economic slowdown.

The contraction of the US economy, as well as its effects on Mexico, began in early 2013. The US Treasury, focusing on the interests of the United States, recommended fiscal measures to protect the US economy and to avoid a spike in unemployment. One of the main recommendations was to increase exports and reduce US imports to invigorate the domestic market (IMF CR/13/334 2013: 35). These measures were expected to shock the Mexican economy, since 80 percent of Mexican manufactured goods were exported to the United States. IMF staff members projected that Mexico would experience a significant drop in manufacturing employment and gross profits if it did not find alternate markets. Mexico was also warned that China was competing for maquila contracts and had offered US corporations a wage guarantee 40 percent lower than Mexico's. The IMF, however, did not expect US corporations to switch labor markets, since Mexico's manufacturing labor force had proved to be reliable and efficient. In particular, Mexican automobile workers had achieved a notable reputation in the manufacturing industry. At that time, Mexico was the second-largest manufacturer of US cars and automotive parts (ibid., 20). Mexico had also set wage-control policies in place to assure US manufacturers that wages would not go up in the near future. China's wages may have been more competitive, but Mexico's close proximity to the United States offset the gains China could offer. Tariffs and transportation costs would more than triple if plants relocated to China.

Although the IMF staff projected that US manufacturing plants would not be relocated, the US Treasury and the IMF advised Mexico to prepare. Raising taxes and creating incentives to attract foreign investments was advisable. This would partly counterbalance the slowdown in the manufacturing sector. President Peña Nieto was also advised to diversify Mexico's banking portfolio because of Spain's banking insolvency problems. Spanish banks owned 11 percent of Mexico's financial institutions, and it might become necessary to replace these investors with more stable financiers (IMF CR/12/317 2012; IMF CR/13/334 2013: 20). To attract new banking investors, Mexico was advised to relax its lending laws. More foreign banks would be interested in investing if they could qualify and offer credit to high-risk applicants. The US Treasury also advised Mexico to raise its treasury bond interest rates to attract American banks. Mexico offered a competitive rate at 4.9 to 6.3 interest, much higher than US rates.

With this economic scenario confronting Mexicans, the president's economic plan was critical to their future welfare. Mexico's middle and upper classes needed a fiscal conservative who would not raise taxes, while the poor needed someone who would not impose new austerity measures.

TAX REFORMS AND PEMEX

A few months after taking office, President Peña Nieto gradually disclosed his reform agenda and began to actively lobby the Mexican Congress for support. His tax reforms were more moderate than expected, while the plans for Pemex were unanticipated. His proposed reforms closely resembled the advice offered by the IMF and the US Treasury. That is, Peña Nieto proposed to enact tax legislation and attract foreign investments to Mexico to counterbalance the effects of the contraction of the US economy. Both recommendations, if implemented, involved the restructuring of Pemex (Revueltas 2012: 3).

In August 2013, Peña Nieto disclosed his tax reforms (Alper and Garcia 2014). He introduced the Fiscal Responsibility Law (FRL), which proposed a series of changes, the most important one being to reduce the amount of taxes paid by Pemex. If Congress enacted this policy, then federal spending would have to be reduced and general taxes raised.

The following month the Mexican Congress passed most of the FRL tax reforms, but set aside the Pemex tax proposal, since Senate representatives did not want to vote on this item until the president unveiled his energy reforms. The public anticipated that the president would make radical changes in Article 27 of the Mexican Constitution. Some speculated that the status quo would be retained, with only minor adjustments to the current oil contracting system, while others projected that Pemex would be privatized (Martin and Rodriguez 2012). With this uncertainty, it was a wise decision for the Senate to delay making any decision about Pemex.

Under the Fiscal Responsibility Law, the Mexican Congress raised taxes affecting all sectors of Mexican society. President Peña Nieto had asked Congress to raise the taxes of Mexico's highest income earners by 12 percent. Congress considered this amount low and instead raised it by 15 percent (Cattan and Martin 2013). Taxes for corporations located along the border were also raised, to 16 percent. These taxes mainly affected US corporations owning maquila plants (IMF CR/13/334 2013: 16).

The public was relieved when Congress did not pass a general flat tax on consumer goods. However, Congress did place a 5 to 8 percent tax on junk food. The main reform affecting most citizens was the new gasoline price policy. By December 2014, all gasoline subsidies were to end. Beginning in 2006, Mexico had gradually raised the price of gasoline a few centavos a month in an attempt to end oil subsidies, which since 1938 had significantly reduced the price of gasoline (IMF CR/6/352 2006: 16, 23; IMF SM/94/41 1994; Manzo Yépez 1996: 31). By the time Congress passed the FRL tax reforms, Mexico's gasoline prices had already been raised considerably and were only a bit lower than US prices. For example, in July 18, 2013, the average price of a gallon of gasoline in the United States was $3.66 and Mexico's prices ranged from US$3.40 to $3.86 (AAA 2014; Pemex 2014b).[17] Mexico City and the states with oil deposits had the lowest prices, whereas the border states had the highest. By raising gas prices, the government hoped to immediately boost oil revenues, because higher domestic prices would cut internal consumption and allow more crude oil to be available for export.

To raise revenue, Congress also adopted the president's banking and telecommunication reforms, which were designed to attract foreign investors. In September 2013, Congress relaxed banking laws, allowing financial institutions to qualify more borrowers for credit (IMF CR/13/334 2013). Banks would be able to qualify high-risk borrowers and charge them significantly higher interest rates. In the telecommunications industry (e.g., satellite transmission cable, internet, cellular phones, television, land lines), foreign corporations were allowed to own 100 percent of a firm, removing the 49 percent limit.

The president did not submit any reforms affecting labor. This was not necessary, since before leaving office, President Calderón had successfully lobbied Congress to pass a series of labor reforms that were designed to attract foreign investments. These reforms were employer-friendly and could be viewed as antagonistic toward labor. Financial caps were placed on class-action lawsuits won by employees. Employment contracts were also modified to allow employers to reduce their workforces without fulfilling employment termination agreements. Likewise, when filling vacancies, employers no longer had to follow a seniority system. New workers or non-senior employees could be hired to fill job vacancies. While these structural reforms were designed to promote financial productivity, raise revenues, and attract foreign investors, the most radical reforms were enacted in the energy field. On December 1, 2013, President Peña Nieto unveiled his energy reforms, including the restructuring of Pemex. He shocked pro-privatization advocates when he announced that Pemex would not be sold.

WERE PEÑA NIETO'S ENERGY REFORMS A CENTRIST POSITION?

President Peña Nieto's energy bill, "Energy Reform: Mexico," addressed the concerns of Mexico's different economic and political sectors by introducing what appeared to be a compromise proposal. To the disappointment of those who favored the privatization of the oil industry, Peña Nieto announced that section 4 of Article 27 would not be nullified. All hydrocarbons, including crude oil, and the subsoil from which such products are extracted remained property of the nation (Estados Unidos Mexicanos, Presidencia de la República 2014a). Interrelated with this pronouncement was the affirmation that the government would continue to have exclusive control over the administration of the oil industry. The most controversial and anticipated revision dealt with section 6 of Article 27, which prohibited private and foreign investments in the exploration, production, and extraction of hydrocarbons (oil, gas, liquids, minerals). President Peña Nieto asked Congress to revise section 6 and allow private firms, including foreign corporations, to invest in these areas. Opening the crude oil market to foreign investors held the utmost significance.

The president's rationale for revising section 6 centered on the premise that seismological data indicated Mexico had 44.5 billion barrels of total crude oil reserves (proven, probable, and possible), which necessitated private investments if deep ocean hydrocarbon reserves were to be extracted (appendix B). In 2011, President Calderón announced that Pemex engineers had discovered crude oil deposits in the Gulf of Mexico. Recent studies now indicated that the deposits were much larger than formerly assessed (González 2013; Iliff 2013; Pemex, *Informe Anual* 2012). If Mexico had sufficient finances, Peña Nieto argued, then Mexico's daily production could rise to 4 million barrels a day from the current level of 2.55 million barrels (Restuccia 2013). Without outside finances, however, Mexico's probable and possible reserves would not be exploited, as recovery from deep reserves required a sizable capital investment that Mexico did not have.

Eleven days after the president announced his energy plan, on December 12, 2013, both the Senate and the Chamber of Deputies (lower house) approved the bill with a comfortable majority (Williams, Martin, and Cattan 2013). The Senate passed the bill with a 95 approval vote, against 28 nays, and the deputy chamber with 354 in favor and 134 against. A series of protests followed throughout Mexico, with PRD lawmakers leading the charge, as well as members of Yo Soy 132, a youth organization that had held many national protests claiming that president Peña Nieto's election was fixed (Excelsior 2013; Gallegos 2013; Restuccia 2013). Andres López Obrador, Pena Nieto's challenger in the presidential election, did not join the marches, as he had suffered a heart attack and was indisposed. The aim of the protesters was to lobby state legislators to vote against the constitutional reform. Before the bill could become law, sixteen out of thirty-one states needed to vote in favor of the legislation (Stevenson 2013).

Opponents identified the most serious problem in the energy legislation to be a lack of clarity in the contract and license provisions. As written, the provisions did not delineate an exact percentage of the profits, risks, or royalties to be shared by the Mexican government and its business partners. The shared-profit agreements could easily be abused. Critics also feared that if Mexico did not place a limit on the amount of investments coming from a single country, then US corporations would take control of the industry. If this happened, Mexico would be left vulnerable when disputes arose, since US corporations have the financial power to organize economic embargos and gain support from the US government. History would be repeated, since the US government has the power to place conditions on the Mexican government during loan or trade agreement negotiations. Opponents also feared that the price of gasoline and heating would skyrocket once foreign corporations owned part of the energy fields (Gallegos 2013). Opponents alleged that foreign corporations would be concerned primarily with making profits, and not with the welfare of the people. They projected that all heating subsidies for the poor would be ended.

On December 16 the Mexican Congress obtained seventeen votes from the states, one more than needed to pass the legislation. Critics throughout Mexico voiced their discontent, arguing that most state legislatures passed the bill without any debate, and that in many cases representatives did not read the entire bill. State legislatures had seventy-two hours to veto or pass the legislation; otherwise, the bill would die.[18] Within Congress, PRD representatives promised to reverse the legislation. Party members announced that by use of popular vote they could win a referendum to rescind the energy reforms (Castillo 2013). Referendums are constitutional means that Mexicans have to reverse government action when the majority of voters disagree with congressional rule. However, PRD representatives chose to wait to hold the referendum until Congress passed the secondary laws governing foreign investments. The PRD anticipated that the laws would allow corporations to receive a much higher percentage of the profit than most people envisioned. They projected that this would anger citizens and energize them to vote in support of the referendum.

Without a doubt, President Peña Nieto's energy reforms succeeded, despite his opponents' threats. The next step for the president and Congress was to design and enact into law the secondary legislation that would govern the procedural implementation of the reforms. The most important secondary laws are those affecting Mexico's crude oil industry, since this is the sector in which foreign companies are expected to invest. Debate for the secondary laws began in the summer of 2014.

THE OIL INDUSTRY REFORMS
AND THE SECONDARY LAWS

President Peña Nieto's energy reforms were widespread and included changes in all energy sectors. The president removed obstacles prohibiting or limiting private investments in the electrical, oil, mineral, and natural gas sectors. In the electrical industry, private investors were also allowed to distribute and market electricity. Due to the extensive nature of the reforms, and to remain within the limits of this study, this discussion will focus on the laws regulating oil exploration and production.

Under the reforms, Pemex lost its monopoly of the petroleum industry, including the exploration, production, and refinement of crude oil. Under sections 4 and 6 of Article 28, Pemex lost its exclusive right to explore, drill, warehouse, and refine hydrocarbons (Estados Unidos Mexicanos, Presidencia de la República 2014a). However, Pemex remained Mexico's premier petroleum agency, with the exclusive right to receive first choice in selecting sites to explore and develop. Pemex also retained control over locations it had previously exploited, and it was not required to enter joint contracts with private investors. It could choose, however, to relinquish property rights over oil fields or enter partnerships with private corporations.

Pemex's primary political function also changed. The agency's fiscal responsibility was no longer solely to generate income to be used for the public welfare of the nation. Under Article 27, section 6, Pemex was converted into a productive state agency, meaning that it was mandated to become a for-profit institution. Its new primary function was to stabilize the peso against the US dollar and ensure that Mexico's GDP did not fall below 4.7 percent (Gobierno de la República de México 2014a: 24). Likewise, as part of Pemex's new mandate, a higher percentage of Pemex profits was to be reinvested in the agency so that it could accrue assets to become a for-profit agency.

The restructuring of Pemex, however, did not mean that the federal budget would no longer receive funds from the sale of oil. On the contrary, the federal government will continue to receive oil revenues, but the sources have changed. Most revenue will come from private investments generated through fees, taxes, royalties, and profit shares attached to the contracts and permits. Under section 6 of Article 27, contracts or licenses will give corporations access to explore and drill for gas and crude oil. Corporations will also be allowed to refine crude oil into petroleum and other liquids.

There are three basic types of contracts: contracts for a set fee, license agreements for leased property, and investment partnerships. Corporations that apply for a set fee contract will be paid for work performed. If oil is discovered, the contract can be renegotiated for a higher fee, depending on the discovery. This policy was previously practiced under President Calderón's administration. In financial lease agreements, the contractor applies for a license. If oil is discovered in the leased property, the profits will be shared based on a prearranged government agreement. If the Mexican government partners with an investor for a profit-sharing or production-sharing contract, both will share the risks and profits. President Peña Nieto chose not to base the partnership system on the piedriega model, since this structure had proved to benefit only investors. Under the old system, the Mexican federal government took all the risks. The investment contract structure was partially designed under the Pemex–Shell Deer Park model, in which the partners shared financial risks if a project failed and profits when it succeeded (Gobierno de la República de México 2014b).

Opponents of the energy reforms proposed that the president did not go far enough to ensure that the government was guaranteed the majority of the profits in the lease agreements and shared partnerships. When President Peña Nieto unveiled his secondary laws platform on April 28, 2014, he tried to re-solve this critique. Under the secondary laws pertaining to hydrocarbons, the president announced that the government would invest a minimum of 20 percent of the cost of a project, and in return would receive a variable profit based on the amount of oil that was extracted and the amount of risk a contractor took (Estados Unidos Mexicanos, Presidencia de la República 2014b: 19). By 2025, when profits from oil production are projected to have increased, the federal government will raise its minimum investment per-centage to 35 percent and acquire a larger percentage of the profit (Gobierno de la República de México 2014a: 35). The variable-share profit percentage will always be in favor of the Mexican government and will be determined by the sitting president in consultation with the Secretaría de Hacienda y Crédito Público (Secretary of Finance and Public Credit) and the Senate (Estados Unidos Mexicanos, Presidencia de la República 2014b: 23). Critics responded that the plan works on a short-term basis but not in the long term (Negraponte 2014). They argued that a shared-profit formula or division of profits should be included in the secondary laws and not determined by the president. If a law is passed establishing a pre-arrangement, investors and lessees do not have the legal basis to litigate when they no longer agree with a long-term contract, whereas if the president has the power to change the agreement, corporations may try to pressure the government to change the contracts (Vera and Farris 2014). Advocates of the president's plan disagreed with this critique and countered that to avoid litigation, protective measures have been set in place in the constitutional articles and in the secondary laws.

On July 19, 2014, the Senate approved President Peña Nieto's twenty-one secondary laws, four of which applied to the oil industry (Negraponte 2014). Debate then moved to the Chamber of Deputies, where the legislation was expected to stall for several months (Reuters 2014). To the astonishment of PRD representatives, who expected to earn the support of the radical sector of the PRI party—to either make significant changes in the legislation or postpone its passage—the majority in the Chamber of Deputies supported the secondary laws within a few days of receiving the bill. On July 28, 2014, the legislation passed with a vote of 330 in support and 129 against (Mendez and Román 2014). The president's secondary laws proposal was enacted on August 11, 2014 (Secretaria de Desarrollo Agrario, Territorial y Urbano 2014). The secondary laws affirm the federal government's exclusive control of the administration of the oil industry. They also stipulate that the Comi-sión Nacional de Hidrocarburos (CNH, Hydrocarbon National Commis-sion) will have control over all technical aspects of the contract and license agreements (i.e., selection of sites, plans, liens, and approval of exploration and drilling plans) (Estados Unidos Mexicanos, Presidencia de la República 2014b). Likewise, the Secretaría de Hacienda y Crédito Público was given full control of all financial matters, and its secretary of finance has the power to design and approve all contracts and licenses pertaining to royalties, fees, and profits.[19] Furthermore, to protect Mexico's assets and clarify section 4 of Article 25 pertaining to investments, the secondary laws clearly reaffirm and stipulate that corporations do not own the oil they extract from Mexican soil. Corporations are solely investors, subject to recover a share of the profits, and they cannot claim ownership over the oil, the soil, or the marketing process. These stipulations allegedly differentiate President Peña Nieto's energy reforms from a privatization model. Under a privatization agreement, investors own the property from which hydrocarbons are extracted as well as the products.

To prevent investors from contesting the shared-profit formula designed by the Mexican government, the secondary laws decree that all arbitration falls under the jurisdiction of Mexican federal law, and no issue can be arbitrated in foreign courts or in a language other than Spanish (ibid.).[20] The secondary laws also stipulate that the executive branch has the power to terminate any contract if the terms are not fulfilled, such as when a contractor, lessee, or investor has failed to commence exploration, has not submitted quarterly reports, has subcontracted or entered a partnership with another company without the approval of the CNH, has caused disastrous ecological environmental pollution, or has intimidated a private landowner to abandon or lease property (ibid., Title 4, chaps. 3–4, pp. 99–105). The executive branch also retains the right to change the boundaries of a concession if the above terms are not met (ibid., 56). This stipulation addresses the concern over long-term contracts if a dispute arises over shared responsibilities.

On July 19, 2014, the Senate also approved the president's tax regime for the oil industry, including Pemex's obligations. The high tax rate imposed on corporations and Pemex revealed why Peña Nieto chose to leave the shared-profit variable open to negotiation. The higher the risk an investor or lessee takes, the more profit a corporation is allowed to collect. For example, corporations will be taxed 70 percent of profits in 2015, 68.75 percent in 2016, 67.5 percent in 2017, and 66.25 percent in 2018. A lower tax rate will be set for all deep ocean projects and determined on a case-by-case basis. Contractors and Pemex will also need to pay a fee for the land under exploitation (1,150 pesos per US$100 per kilometer for the first 60 months, and 2,750 pesos per $200 after 61 months) and a variable fee for each barrel of crude oil, which is dependent on the market value of oil (SCHP 2014). For investors who entered a shared-profit arrangement, and also in the case of Pemex, they can subtract the operation expenses from the taxes owed. This does not apply to contractors who entered a license agreement.

Two days after the tax regime was announced, the director of Pemex, Emilio Lozoya, began accepting bids for exploration in regions where large probable deep ocean deposits have been identified (e.g., Poza Rica Altamira, Burgos, Chicontepec) (Pemex 2014c). These are high-risk investment zones where large financial gains can be made.

After the president's secondary laws proposal was accepted by the Senate, the PRD movement to rescind the legislation continued and temporarily gained momentum. Peña Nieto's staff had unsuccessfully tried to stop the referendum after appealing to the Chamber of Deputies. The legal argument raised was that Mexican law permits referendums to be held only when the challenge does not apply to constitutional decrees. On April 25, 2014, the deputy chambers disagreed and consented to the petition drive (Nieto and Jiménez 2014). Within days, members of the PRD obtained over 4 million signatures, well in excess of the 1.6 million signatures needed. Immediately, the President's Office appealed to the Mexican Supreme Court. On October 24, 2014, the justices ruled in a 9 to 1 decision to support the president's appeal. They offered the opinion that the Mexican Constitution was not violated (TeleSUR 2014). The decision was based on the premise that tax and federal spending legislation cannot be repealed through a referendum vote. The adverse ruling quickly ended the aspiration of PRD activists to pursue their complaints through the courts and Congress. The only alternative for modifying Article 27 is for the people to elect a president and a Congress that support restoring the nationalization of Mexico's oil industry. In retrospect, President Peña Nieto was well prepared for the national protest and astutely appended tax law reforms to the energy laws, thus establishing the legal basis to prohibit citizens from intervening.

A NEW ERA, OR WILL HISTORY REPEAT ITSELF?

The question that remains is, who benefits from the reforms? At this point in time it is difficult to project what will transpire. It is uncertain whether the terms of the contracts and licenses are sufficiently attractive to convince US corporations to partner with Mexico. That Mexico is the sole owner of the crude oil, as well as of the subsoil where the deposits lie, may discourage companies from taking the risk to explore for oil. President Peña Nieto's protective laws, which give the executive branch the power to terminate contracts and change exploration boundaries, may also be a concern for foreign corporations. If foreign companies do not accept the current terms, then the Mexican Congress may need to revisit the energy reforms and consider making oil exploration in Mexico a more lucrative business venture and a less regulated enterprise.

For Mexicans, if the president's plans do become a reality, then increased revenues will stimulate the economy and benefit the entire nation as a whole. Mexican citizens, however, need to remain vigilant as the reforms unfold. If US companies invest in Mexico and exploration more than doubles, as projected by the Mexican government, chapter 4 of title 4 of the secondary laws (Del Uso Y Ocupacion superficial) gives the government the right to expropriate the property where companies identify oil deposits (Estados Unidos Mexicanos, Presidencia de la República 2014b: 100–105). The Mexican government has adopted laws to protect landowners from the predatory actions of foreign corporations, yet at the same time it has placed few restrictions on its own power.

To protect Mexican private landowners from private corporations, the Agricultural Judiciary Institute has been mandated to regulate surface soil private leases and offer an assessment of the amount of rent the companies should pay. This requirement has been instituted to ensure that foreign corporations pay a fair fee for the rented surface land, as well as reimburse landowners for property damages after the rental contract expires. The institute will also investigate whether landowners are being intimidated when negotiating a rental fee, or if they are being coerced to sell their property. This regulatory policy was designed to avoid past practices. As discussed in chapter 2, during the Porfirian period the Mexican government allowed US and British companies to use the Denunciation Law of May 26, 1894, to force people off their land by the use of violence and intimidation. Once people left, the Mexican government declared the land abandoned and transferred title to the companies. The rationale was that US and British corporations would make the land productive. To prevent such events from being repeated, the institute is empowered to rescind companies' contracts if it confirms allegations of coercion. Certainly, these are reasonable protective policies. But the Mexican government has also decreed that if oil is found on private property, the government will expropriate the land if the proprietor refuses to sign a rental contract with a company. This latter point gives Mexicans few legal protections and is reminiscent of the Porfirio days, when common people had few options or legal rights.[21]

For the US government, the restructuring of Pemex is a winning proposition all the way around. If President Peña Nieto's plans unfold as projected, then Mexico's economy will prosper and lead to increasing trade with the United States. A prosperous Mexico also means that new jobs will be created and fewer Mexicans will need to emigrate to the United States in search of employment. US national security interests will also benefit if a positive scenario unfolds. Mexico, as an ally nation of the United States, will be able to increase its crude oil production and have more oil available for sale. But if Peña Nieto's plans fail and political turmoil follows, this will not necessarily lead to the exodus of US petroleum corporations. By then, US corporations would have reentered Mexico and regained the legal standing to remain there.[22]

Asymmetrical Codependency

A FUNCTIONAL CAPITALIST RELATIONSHIP

*I*N THIS BOOK, I HAVE EXAMINED FARM LABOR AND crude oil as two sources of energy that bind the economies of the United States and Mexico. Historically, for US agriculture to function efficiently, it has required a continuous and constant flow of agricultural labor from Mexico. At the same time, the US government has relied on scheduled imports of Mexican crude oil to fill its consumption demand. Mexico, in turn, depends on the finances that these energy sources generate for the Mexican economy. Throughout my analysis, the concept of codependency has been used to characterize this association. I have argued, however, that this relationship is asymmetrical, because the US government historically has had the financial power to shape the direction of this interdependency. Mexico has accepted this asymmetry because it is dependent on the financial assistance it receives during periods of economic stress. This economic dependency is complex, however, because Mexico's economic stability is important to the United States. A ruined Mexican economy can lead to increased migration and the closure of Mexico's import markets. Currently, Mexico is the United States' second-largest export market and third-largest trading partner (US Department of State 2014). Furthermore, Mexico's stability is particularly important for the US agricultural industry, since it is the third-largest importer of US agricultural commodities (USDA 2015c). Only China and Canada exceed Mexico's agricultural imports.[1]

This book has attempted to expand dependency theory from its traditional perspective that weak countries are dominated by powerful nation-states. Without dismantling this thesis, the historical analysis advanced in these chapters has illustrated how a wealthy nation can also become economically dependent on the resources of a country in which it exerts significant control. In assessing US-Mexican relations over farm labor and crude oil, I showed that the agreements enacted by both nations reflect a capitalist logic. That is, both nations share the philosophical principle that international commerce must be promoted and supported. The flow of commodities across the border must be regulated, but not obstructed by tariffs or policies that impede free market competition.

In this book, it has been shown that while crude oil accords have been strategically crafted and monitored by the governments of the United States and Mexico, the same type of oversight has not been given to regulating the flow of Mexican farm labor. This form of energy, that is, the flow of agricultural labor across the US-Mexican border, has been neglected and exploited. During the bracero years, the US government periodically refused to regulate the program and turned its management over to the agribusiness industry, resulting in wage and housing abuses. Under a capitalist system, it is logical such abuses would take place because agribusiness employers will try to maximize their profits by reducing the cost of labor. What is uncertain, however, is why over the years the Mexican government has failed to protect the exchange of this human commodity in the same manner that it forcefully protected its oil trade. Once again, a similar question can be raised about the treatment of Mexican farmers in the aftermath of NAFTA. What is the capitalist logic that leads the Mexican government to neglect the livelihood of its small- to mid-scale farmers? Why were agricultural free trade agreements allowed to continue when it became obvious that they had negative impacts on the Mexican rural sector?

The responses to these questions cannot be attributed solely to a theoretical economic miscalculation. When they were negotiating the NAFTA agreement, bureaucrats must have known that there was a great probability that the manufacturing industry would be unable to absorb the displaced agricultural workers. Perhaps this was a conscious decision made by Mexican bureaucrats based on the callous projection that when people need to eat, they will cross the border in search of employment. As discussed in chapter 5, the number of manufacturing plants established after NAFTA did increase, and their efficiency greatly improved, yet this infrastructure did not lead to the massive expansion of the workforce. The maquila system thrived because of the fast production output per worker and not because more jobs were created. Within a decade of the passage of NAFTA, when it became obvious that manufacturing employment was not expanding fast enough to counteract the displacement of the agricultural population, it was unfortunate that the government did not set in place other mechanisms to ensure that such people found other forms of employment that paid a living wage.

What followed was the movement of a large percentage of the displaced farmers into the undocumented labor stream of the United States. They joined the millions of unemployed Mexicans who had lost their livelihood because of the disastrous economic events that had transpired. But agricultural workers, unlike other Mexican workers, had become an important infrastructural component of the American economy, and their labor was welcomed because most American citizens were unwilling to perform farm labor. Thus it is not illogical to deduce that Mexican bureaucrats failed to develop an alternative plan to find employment for displaced Mexican farmers because they projected that the US agricultural industry would absorb them.

At this moment, the absorption of Mexican agricultural workers into the US economy appears to be the ramification of a capitalist supply-and-demand labor reality, rather than a colonial relationship. As has been shown in this book, Mexico is a sovereign nation that historically has practiced a capitalist system. It is my contention that the flow of people across the US-Mexican border is currently a component of the Mexican capitalist structure. The outmigration of Mexicans is a significant structural conduit that allows Mexican elites to ignore income inequality and callously expect that Mexico's unemployed will find work in the United States. In evaluating this escape-safety valve mechanism used for generations by the Mexican government, I propose that as long as immigrants are allowed to pledge their political loyalty to their new nation, Mexicans cannot be considered a people who are fleeing a colony. They are instead people who are part of a free labor market and in search of the highest wages. They are also persons whose nation of origin did little to keep them at home. On the other hand, if this naturalization structure ends and Mexican immigrants are no longer allowed to become US citizens, then Mexico would have been transformed into a colony that solely supplies cheap labor. Mexican migration to the United States would no longer be part of a free labor market movement, where the poor and the weak find a forum of social mobility through industriousness, competition, and the exchange of labor for a fair living wage. The mutation of this political structure would certainly reflect the core-periphery model introduced by Raúl Prebisch and popularized by Immanuel Wallerstein.

The same capitalist logic applies to the management of crude oil. As long as the Mexican government retains control of management, ownership, regulation, and most of the crude oil profits, Mexico cannot be considered a colony. However, if Mexico privatizes the industry and loses control of it to US corporations, then this will indicate that the nation is undergoing a fundamental transformation in which its most valuable resource is now owned by another nation. Upcoming events in the Mexican oil industry will provide a window into Mexico's future relationship with the United States. Will foreign corporations accept Mexico's investment laws and share the liabilities, risk, and profits with Mexico, or will they refuse to invest unless the Mexican government takes most of the financial risks and allows arbitration disputes to be litigated in foreign courts? President Peña Nieto has affirmed that all arbitration disputes will be litigated in Mexican courts and be governed by Mexican federal law. The question is whether this position will be altered in the future. If the Mexican Congress allows oil disputes to be litigated in the judicial court of the investor's nation, then Mexico will have forfeited control over the oil industry and begun to turn its codependent sovereignty into a type of dependency characteristic of a colony.

This book finds that the typical traits associated with a colony do not characterize US-Mexican relations. The relationship is better understood as asymmetrical codependency, that is, a relationship that benefits both nations because they are capitalist societies that value free market competition, the accumulation of wealth through reduction of the cost of labor, limited government intervention in the markets, and the philosophy that citizens must be responsible for themselves.

In retrospect, both nations are content with their arrangements regarding farm labor and crude oil because this energy structure is mutually beneficial and profitable. For example, Mexico has the power to enact infrastructural changes in its public budget to reduce the occupational displacement of small- and mid-scale farmers, yet it chooses not to. The continuous flow of Mexican farm labor to US farms has become a fixed structural component of Mexico's economy. This structure could be terminated or lessened by instituting protective agricultural tariffs that allow domestic farmers to better compete in the global market, yet the Mexican government prefers to maintain the status quo.

Similarly, the US government could certainly gain its independence from a codependent relationship and end its demand for Mexican agricultural labor. At any time, the US Congress has the power to import workers from other nations or terminate the flow of agricultural workers, but it chooses not to. If the system is not broken, why change this silent arrangement? It is not cost-effective to terminate a structure that ensures a steady flow of vulnerable agricultural workers who are given few protections when they arrive in the United States.

A similar argument can be made about US-Mexican crude oil trade. The US government and US corporations have access to many oil markets around the world and could easily enact an embargo against Mexico if they chose to do so. Let us not forget that such an event did occur on the eve of World War II, after Mexican president Lázaro Cárdenas enacted the expropriation decree of 1938. Without a doubt, the US government has the power to terminate its dependence on Mexican crude oil, but it chooses not to. Retaining an amicable crude oil trade alliance has benefited both nations: Mexico has a dependable customer, while the US government has a dependable supplier across the border.[2] As long as Mexico continues to produce large quantities of oil and remains a source of energy, this relationship is beneficial to the governments of both nations.

Pemex Assessment of Mexico's Proven Crude Oil Reserves, 1976 to 2014 (in millions of barrels)

Year	Proven Reserves	Historical Event
1976	7,279	
1977	10,428	
1978	28,407	
1979	30,616	
1980	44,161	
1981	48,084	
1982	48,084	
1983	49,911	
1984	49,260	
1985	48,612	
1986	48,612	
1987	48,041	
1988	47,176	
1989	46,191	
1990[1]	45,250	
1991	44,560	
1992	44,292	
1993	44,439	
1994	44,043	
1995[2]	43,127	
1996	42,146	
1997	42,072	Reserve recalculation discussions
1998	41,392	
1999	24,916.6	Recalculation of reserves begins using US Securities and Exchange Commission definition
2000	25,070.4	

Year	Proven Reserves	Historical Event
2001	23,525.4	
2002	21,892.7	Reserve recalculation discussions
2003	20,077.3	US Securities and Exchange Commission definitions used from here on
2004	18,895.2	
2005	17,649.8	
2006	16,469.6	
2007	15,514.2	
2008	14,717.2	
2009	14,307.7	
2010	13,992.1	
2011	13,796.0	
2012	13,810.3	
2013	13,868.3	
2014	13,438.5	

Sources: Most recent estimates used and based on proven crude oil equivalent. Only 1976–1978 include condensate crude oil. Pemex, *Statistical Yearbook 1988* (Exploration and Production section) for 1976–1978; *Statistical Yearbook 1990* for 1979–1985; *Statistical Yearbook 1996* for 1986–1988; *Statistical Yearbook 1999* for 1989–1998; *Statistical Yearbook 2009* for 1999 revised; *Statistical Yearbook 2012* for 2000–2011; *Statistical Yearbook 2013* for 2012–2013; *Statistical Yearbook 2014*; Pemex 2014d; Iliff 2013.

[1] Pemex recalculated the 1980s proven reserves in 1990 and 1995. 1980s yearbooks have higher estimates.

[2] In 1997 the US Securities and Exchange Commission advised Mexico to recalculate the procedure for proven reserves. In 2002 the SEC once again set new standards that Mexico followed. They were implemented in 2002, leading to the recalculation of the 2000–2001 estimates (Pemex, *Informe Anual 2003*: 55; Shields 2003).

Pemex: Total Crude Oil Reserve Estimates, 2003 to 2014 (in billions of barrels)

Year	Total[1]	Proven	Probable	Possible
2003	50.032	20.077	16.965	12.990
2004	48.041	18.895	16.005	13.141
2005	46.914	17.650	15.836	13.428
2006	46.418	16.470	15.789	14.159
2007	45.376	15.514	15.257	14.605
2008	44.483	14.717	15.144	14.621
2009	43.563	14.308	14.517	14.738
2010	43.075	13.992	14.237	14.846
2011	43.074	13.796	15.013	14.265
2012	43.837	13.810	12.353	17.674
2013	44.530	13.868[2]	12.306	18.356
2014	42.158	13.439	11.377	17.343

Source: Pemex, *Statistical Yearbook 2013*: 13; Pemex 2014d.

[1]Crude oil equivalent includes crude oil and condensates. Condensates are plant liquids and gas that are obtained when drilling crude oil. All figures have been rounded.

[2]In March 2013 Mexican president Enrique Peña Nieto announced that the proven reserves had reached 13.870 billion barrels and the crude oil equivalent 43.840 billion barrels (Iliff 2013).

Notes

1. The economic classification of countries in social science literature advances competing terminologies that are similar and interchangeable (see Nielsen 2011). Common terms are: "poor/rich," "backward/advanced," "underdeveloped/developed," "developed/less developed," "North/South," "latecomers/pioneers," "colonizer/colonized," "third world/first world," and "industrialized/developing."

In the years after the Second World War, Mexico has been described as a developing country, a developing oil-nation, and an emerging and middle-income nation. The term "third world" has also been applied to Mexico, in reference to its legacy as a Latin American Spanish colony (Frank 1967; Meyer, Sherman, and Deeds 2007).

Over the years, within the IMF and the World Bank, the term "developing country" underwent various definition shifts (Nielsen 2011: 7). In the early 1970s, the developing category included all countries that were not industrialized or did not have advanced economies. Within a few years, due to the monetary value oil came to hold in the international markets, the category came to be redefined. Developing nations were grouped as developing oil nations and developing non-oil nations (see IMF *Annual Report* 1982). Developing oil nations were characterized as those undergoing widespread industrialization and with advanced technologies. The standard of living in these nations, however, was concluded to be uneven, with wealth being concentrated among the top 10 percent of the population, while the majority lived in or near poverty and did not have access to advanced technologies. Developing nations without oil resources were characterized as countries that mainly relied on subsistent farming, had little industry or advanced technology, experienced high unemployment and a low standard of living with the majority of the population living under the poverty level, lacked basic infrastructure in most regions, and had no GDP growth. In 1989 the oil-grouping variable was no longer used as the main criterion distinguishing developing countries, since some of the non-oil-producing nations were economically better off (Nielsen 2011). Nations without oil, such as India, had an advanced manufacturing sector and a growing GDP, while some of the oil-producing nations, such as Mexico, were highly indebted and had little or no GDP growth.

In 2010 the United Nations, the International Monetary Fund (IMF), and the World Bank revised and adopted common terms to classify the economies of the world. A nation's Gross National Income (GNI) became the main classificatory criterion (Nielsen 2011: 16). The two main country categories are "developed" and "developing." The term "developed" replaced the popular category "industrial country," which referred to first world countries. Interchangeable terms for a developed country are "high-income country" and "advanced country." Examples of developed countries are the United States, the United Kingdom, Germany, Japan, and France.

Developing countries are subdivided into two groupings: "developing low-income" and "developing middle-income." A middle-income nation is also referred to as an "emerging economy." The placement of a developing nation within this subgrouping is based upon its GNI index, with consideration of the life expectancy of its population and the number of schooling years completed per person.

2. Alexander Lesser (1933) was among the few anthropologists who offered a critical analysis of colonialism. His research was marginal to the field of anthropology and focused on Native American modes of resistance against Anglo American attempts to dehumanize nonwhites (see Mintz 1985a).

3. By the late 1940s, a few anthropologists advanced developmental theories that were more complex. Writing within a critical anthropological tradition, Leslie White rerouted the debate by applying Marxist stratification principles to the developmental framework. In 1949, White wrote *The Science of Culture: A Study of Man and Civilization*, in which he argued that societies across the world reached different levels of technological development largely due to their organization of labor, the government they practiced, their harnessing of energy, and their land tenure practices. The prosperity or stagnation of a society or nation depended upon how these societal conditions were assembled. Although White's analysis paralleled Herskovits's developmental continuum (backward/advanced), it was radically different because a utopian final stage would not arise after a society was colonized. Controlling and exploiting the markets of the less-advanced societies was central to the livelihood of the dominant nation. Over the years, the social evolution of an advanced nation depended on its ability to monopolize new markets and use its dominance to prosper.

4. See Shipway 2008 for a historical review of the restructuring of US, British, and French colonies after the end of the Second World War.

5. Written in 1969 and translated into English in 1979, the book *Dependency and Development in Latin America*, by Fernando Henrique Cardoso and Enzo Faletto, advanced a core-periphery historical analysis of Latin America's economic development.

6. During the 1980s, the writings of anthropologists contributed to the prominence of the core-periphery model within academic circles. Eric Wolf (1982) and Sidney Mintz (1985b) advanced provocative historical studies of how Europe and the United States came to control the economies of many countries in Latin America and Africa. Their writings, however, focused on how economic and ideological structures of exploitation were formed and not on how these structures developed into the pres-

ent. Both scholars were highly influenced by the writings of Frantz Fanon, Edward Said, and Albert Memmi.

CHAPTER 2

1. For an extended discussion of US citizenship laws in the annexed Mexican territories, see Menchaca (2011, 2001) and Weber (1982).

2. In 1900, Edward Doheny established the Mexican Petroleum Company with other investors. A year later, to become independent and in full control of his operations, Doheny established the Huasteca Petroleum Company (Brown 1993: 27, 38).

3. Under Article 23 of the Mexican Constitution, the Mexican Congress recognized the right of capitalism to exist in Mexico. Mexico's labor laws were to be designed to provide a legal framework in support of harmonizing relations between capital and labor (Santiago 2006: 241).

4. Estimates of the deportations range from 400,000 to 750,000 (see Hoffman 1974). The US government did not maintain a precise count of the number of people deported. Scholars estimate the size of the deportee population based on government data and news reports. A general agreement found in the literature is that a large percentage of the deportees were the American-born children of Mexican immigrants. Since the US Census reported that in 1920 only 486,418 foreign-born Mexicans resided in the United States, one should consider the census data when evaluating the estimates (see Menchaca 2011: 216, 253).

5. In 1934 Mexico established a state-owned oil company called Petromex and at that time acquired the expertise to run different functions of the oil industry (Meyer 1972: 197).

CHAPTER 3

1. In 1910, 85 million people resided in farms, and in 1963 the number fell to 20 million (US Department of Labor 1964a: 9).

2. See Galarza 1977 for a discussion of the worst transportation accident, involving braceros and farmworkers. In the 1960s it was common for employers to transport their workers to the fields in improperly renovated buses. Flatbed trucks were converted into buses with no seats or compartments separating tools and machetes from the passengers. On September 17, 1963, field workers sitting on the bed of a converted bus were unable to see that an approaching train was about to collide with them since the hood cover had no windows. The train killed or injured fifty-eight of the passengers. Thirty-two were braceros who were dismembered and crushed to death by the train and the tools inside the truck. In 1964 Galarza wrote a seventy-two-page report for the House Education and Labor Committee of Congress in which he documented the dangerous transportation conditions that braceros faced. Galarza was a labor activist and held a doctoral degree in economics from Columbia University.

3. In the early 1970s the Teamsters Union competed with the UFW for union contracts and often negotiated sweetheart deals with farm employers (Taylor 1975). The Teamsters Union was primarily concerned with improving the working conditions of agricultural employees who worked in mechanized occupations, such as machinery operators, truckers, and assembly line packers. Teamster representatives negotiated higher wages and better work conditions for those working in mechanized occupations, while doing little for the common farmworker. For the growers, this reduced farming costs, because the majority of the workers were crop pickers.

4. In 2000 the US Department of Labor estimated that the agricultural labor force numbered between 1,374,000 and 3,382,000 (Levine 2004b: 10; US Department of Labor 2002: Table 1).

CHAPTER 4

1. Cantarell deposits were discovered in 1976 but were concluded to be massive in 1978.

2. By 1980, Mexico's annual gross domestic product grew to 6.7 percent, up from 4.23 percent in 1976 (Flores-Quiroga 1998: 159; INEGI 2000b: 43).

3. In 1981 the US federal minimum wage was $3.35 an hour (Marshall 2013). In that year, the federal minimum wage was extended to farmworkers and wages began to rise. In some states, wages surged, such as in California, where large-scale growers reported severe labor shortages and began paying from $4.25 to $7.17 (Employment Development Department Research Division 1986; Ventura County Agricultural Association 1986).

4. Foreign direct investment in Mexico accelerated after 1991. Total foreign investments in 1990 were US$35.5 billion, and were $67.8 billion by 1992 (Chávez Ramírez 1996: 145); by January 1994 they had grown to $102 billion (GAO 1996: 41).

5. In 1991, foreign corporations were permitted to own 30 percent of a Mexican bank (Bancomer 2004; IMF CR/9/7 2009).

6. In 1990, Mexico's food subsidy program CONASUPO was nearly eliminated (IMF EBS/90/10 1990).

7. During Mexico's economic crisis of the 1980s, SAM was replaced by PRONASOL in 1986. SAM had been a large-scale subsidy farm program to assist farmers of all income levels. PRONASOL was instituted to provide financial assistance only to farmers who produced specialty crops, such as corn and beans. In 1993 PROCAMPO replaced PRONASOL and reduced government subsidies further. PROCAMPO aid was given only to large-scale agricultural corporations producing for market (Fox 1992; Guerrero Andrade 2005; Schatan 1987).

8. The number of people reporting agriculture as their primary occupation in 2012 was 13.6 percent (INEGI 2012b: 16).

9. From 1995 to 2000, manufactured exports increased by 80 percent (Kose, Meredith, and Towe 2004: 15), and by 163.8 percent in 2009 (Domínguez and Fer-

nández de Castro 2009: 113). Although these statistics confirm that Mexico's manufacturing exports have increased over the years, the profits do not necessarily stay in Mexico, since most companies are foreign-owned (INEGI 2000b: 115).

10. INEGI reported that in 2011, Mexican industrial employment totaled 23.3 percent of the total employed (INEGI 2012b: 16).

11. In Mexico, corn and wheat production declined after NAFTA. In 1994, 8.2 million hectares of corn were harvested; by 1999, production had fallen to 7.2 million hectares (INEGI 2000a: 19). A similar decline occurred in the wheat industry, where 1.1 million hectares were harvested in 1989 but only 679,999 hectares in 1997 (Avila Dorantes, Cortes, Rinderman, and Palacio Muñoz 2001: 92).

12. Under article 704 of NAFTA, corporations were prohibited from offering subsidies to trading partners (International Institute of Sustainable Development 2007). Governments were also discouraged from offering home corporations subsidies, but were not prohibited. They were encouraged to move toward a system where domestic support measures would have minimal impact on the value of trading commodities. In the area of agriculture, the Mexican government nearly ceased giving subsidies to farmers, while the US government did not. The US corn industry received the largest percentage of the agricultural subsidies. In 1999, the subsidies amounted to US$7.3 billion, and had increased to $9.5 billion by 2005 (Relinger 2010; USDA 2015a).

13. On May 16, 2006, the Foreign Relations Committee of the US Senate recommended that the State Department encourage countries with significant underdeveloped oil reserves to allow US investments in their oil industry, in order to increase the available supply of oil for the United States (US Senate Foreign Relations Committee 2006: 25).

14. The congressional bills related to agricultural labor were 109th Congress (2005–2006), 3 bills; 110th Congress (2007–2008), 6 bills; 111th Congress (2009–2010), 4 bills; 112th Congress (2011–2012), 2 bills; and 113th Congress (2013–2014), 2 bills (www.thomas.gov).

CHAPTER 5

1. In 2004, US residents began to reduce their consumption of oil, which in the last twenty-five years had ranged on an annual basis from 25 to 27 percent of world consumption (EIA 2013b, 2015a). By 2020, China's consumption is expected to have exceeded that of the United States (IEA 2015).

2. In 1973, Mexican president Luis Echeverría was hostile toward OPEC (see Díaz Serrano 1992; Nasar 1981: 128, 186).

3. Source: EIA 2013c.

4. During the Baker Plan negotiations, Mexico agreed to an oil price contingency agreement called the "Stand-by Arrangement of November 20, 1986." The IMF guaranteed banks 50 percent of the amount loaned to Mexico if the banks lent

new money. In return, the Mexican government agreed to an oil price contingency agreement in which Mexico would not be able to borrow new money if a barrel of oil was over US$14; if the price of oil fell below $9, the IMF would give Mexico financial assistance. The intent of the agreement was for the IMF to closely monitor Mexico's oil sales (Boughton 2001: 429, 437). This was the first time that Mexico conceded to an oil monitoring agreement.

5. For most years, the US Energy Information Administration (EIA) and Pemex advance similar calculations of Mexico's proven crude oil reserves. From the late 1980s to 2003, Mexico's estimates are generally lower. In 1989 and 1990, however, the EIA and Pemex estimates were very different. Pemex reported a decrease of its reserves, while the EIA reported an increase. The different tabulations may stem from Mexico not including in its calculations the crude oil that had been presold to the United States under the oil exchange agreement of 1990.

6. Under President Salinas's administration in 1992, petrochemicals, which were derived from crude oil, were redefined as chemical compounds not under the restrictions of the Mexican Constitution. This legally allowed the government to encourage private investment in petrochemical production (Chávez Ramírez 1996: 108; Manzo Yépez 1996: 121). Compañia Mexicana de Estaciones de Servicio (Mexican Company of Service Stations, CODESSA) was also dissolved in 1992. CODESSA was an agency under Pemex mandated to manage the state-owned gasoline stations. Beginning in the mid-1980s, state-owned gas stations were gradually sold to the public, and by 1992 they were completely privatized. All gas stations, however, were required to sell only Pemex-refined gasoline (Flores-Macias 2010).

7. See Puyana Mutis 2006 for a comparative analysis of the profits and costs involved in managing a private or state-owned agency.

8. The IMF reported that in the early 1990s, most long-term bonds and notes were sold at a 14 to 26 percent interest rate, while INEGI reports that by 1994, before the economic crisis, some were selling at as high as 37 percent (IMF SM/94/41 1994: 46, 59; INEGI 1996: 163).

9. The loans of 1984, 1987, 1989, and 1990 were to be paid by 2004. In 2013 they continued to be active loans (US Department of State 2013: 192–193).

10. Mexican billionaire Carlos Slim was the wealthiest person in the world in 2007 and from 2010 to 2013 (see www.forbes.com).

11. The World Bank states that in 2012, Mexico's poverty rate was 52.3 percent, while IMF states 51.3 percent.

12. In 2013, Mexico had a total foreign debt of US$372.6 billion, with US corporations owning 33 percent of Mexico's debt ($270.5 billion is the government's public debt, and the rest is private debt; IMF CR/13/334 2013: 35).

13. The size of the Mexican upper class remained relatively unchanged, at around 6.6 percent of the total population (*Forbes Mexico* 2014).

14. The Mexican middle class is characterized as having adequate housing, defined as a home with the necessary number of rooms to comfortably accommodate a family (INEGI 2013b). Members of the middle class have adequate clothing and

sufficient food, they own a computer, and they qualify for credit (Aristegui Noticias 2013; O'Neil 2013).

15. Mexico's illiteracy rate declined from 12.4 percent of the total population in 1990 to 6.4 percent in 2011 (INEGI 2012b: 10).

16. From 2000 to 2009, the number of primary schools actually declined, from 99,008 to 98,575, when the school-age population ages 5 to 14 increased by approximately 3 percent (INEGI 2009a: 85, 2014: data online).

17. On July 18, 2014, Mexico's gasoline prices surpassed US prices. The average price for gasoline in the United States was US$3.58 a gallon and in Mexico it ranged from US$3.93 to $4.50 (Pemex 2014b).

18. By December 18, 2013, twenty-four legislatures had approved the energy reforms (Estados Unidos Mexicanos, Presidencia de la República 2014b: 2).

19. Sections 5 and 8 of Article 25 of the Mexican Constitution pertain to the regulatory process. The secretary of the Department of Energy is responsible for interpreting policy designed by Congress (Gobierno de la República de México 2014a: 17). All financial matters associated with contracts and licenses will be negotiated and finalized by the Secretaría de Hacienda y Crédito Público. In turn, La Comisión Nacional de Hidrocarburos (CNH), a commission composed of seven members, will evaluate the secretary of energy's administration of the petroleum and electrical industries. The commissioners will hold judicial authority to confirm the contracts and licenses. The president appoints the commissioners, and the appointments are ratified by two-thirds of the Senate. All agencies are accountable to the president.

20. During the 2008 constitutional debates over Article 27, many senators did not oppose the agreement that judicial arbitration take place in foreign courts when disputes arise over oil contract agreements (Barajas Duran 2008; Perez 2010).

21. The secondary laws stipulate that gasoline prices will continue to rise on a monthly basis from 2015 to 2019 until the value of gasoline reaches a competitive market level. Afterward, prices will be adjusted based on Mexico's inflation level (Estados Unidos Mexicanos, Presidencia de la República 2014b: 38). Price hikes have been instituted because the Mexican government purchases a large percentage of its gasoline from abroad; Pemex does not have sufficient refineries to convert crude oil into gasoline. Mexican citizens should be vigilant on this issue, since current Mexican gasoline oil prices have surpassed US gasoline prices. Gasoline prices could become a serious problem for people living in poverty, since wages in Mexico are very low in comparison to those in the United States. Mexico's minimum wage in Zone A, Mexico City, is 64.76 pesos, or US$4.88, a day, with other geographic zones slightly lower (Presidencia de la República 2014).

22. The secondary laws have not addressed the legal rights that contractors have over the transfer of hydrocarbons (Vera and Farris 2014). Companies may have the right to levy a transfer fee when they turn over the oil to the Mexican government.

CONCLUSION

1. In 2013, Mexico imported US$18.1 billion in agricultural commodities from the United States; from China, $25.5 billion; from Canada, $21.4 billion; and from the European Union, $11.8 billion (USDA 2015c). Other Latin American countries import a small percentage of US agricultural products at costs ranging from $1.9 billion to under $159 million.

2. In November 2014, the global price of oil fell to around US$48.43 a barrel when OPEC refused to reduce production and flooded the market. Mexico was one of the few oil-producing countries that did not have to worry about the massive drop in the price of oil because it had contracts guaranteeing the price of a barrel of oil at US$76.50 (Arnsdorf 2015; Caruso-Cabrera 2014). The price of oil rebounded by May 2015 at $60 a barrel, and its future value is dependent on global politics (Zhou 2015).

Bibliography

Aguirre Beltrán, Gonzalo. 1991. *Formas de gobierno Indigena.* 3d ed. Obra antropológica 4. Mexico City: Instituto Nacional Indigenista, Universidad Veracruzana.

Al-Chalabi, Fadhil. 1984. *La OPEP y el precio internacional del petróleo el cambio estructural.* Mexico City: XXI Editores.

Alonso, Austin. 2013. "U.S. Ag Secretary Says Immigration System Stunts Industry Growth." *Kansas City Business Journal,* June 21, 2013. http://www.bizjournals.com/kansascity/news/2013/06/21/secretary-vilsack-immigration-reform.html?page=all.

Alper, Alexander, and David Alire Garcia. 2014. "Mexico Energy Reform Won't Force Pemex into Contracts—Lawmakers." *Reuters Edition,* February 19, 2014. http://www.reuters.com/article/2014/02/20/mexico-reforms-oil-idUSL2NoLO 28R20140220.

American Automobile Association. 2014. "National Average Prices, July 18, 2014." http://fuelgaugereport.com.

American Farm Bureau Federation. 2015. "Agricultural Labor Reform." Washington, DC: American Farm Bureau Federation. http://www.fb.org/index.php?action=issues.aglabor.bills.

Andrade, Armando, and Nicole Blanc. 1987. "SAM's Cost and Impact on Production." In *Food Policy in Mexico: The Search for Self-Sufficiency,* ed. James Austin and Gustavo Esteva, 215–248. Ithaca, NY: Cornell University Press.

Appadurai, Arjun. 2006. *Fear of Small Numbers: An Essay on the Geography of Anger.* NC: Duke University Press.

———. 2000. "Grassroots Globalization and the Research Imagination." *Public Culture: Society for Transnational Cultural Studies* 12(1): 1–19.

Aristegui Noticias. 2013. "Diferencias entre clases baja, media y alta en México: Inegi en CNN." June 16. http://Aristeguinoticias.com/1706/mexico/diferencias-entre -clase-baja-y-media-inegi-en-cnn.

Arnsdorf, Isaac. 2015. "The Oil Rally Looks Doomed, in Five Charts." *Bloomberg Business,* April 28. http://www.bloomberg.com/news/articles/2015-04-28/the-oil -rally-looks-doomed-in-five-charts?bpop=79127812.

Associated Press. 2006. "Despite Smiles, Divisions Linger after U.S.-Mexico-

Canada Summit." *USA Today*, March 31. http://www.usatoday30.usatoday.com/news/.../2006–03-31-bush_x.htm.

Austin, James, and Gustavo Esteva. 1987. *Food Policy in Mexico: The Search for Self-Sufficiency*. Ithaca, NY: Cornell University Press.

Avila Dorantes, Jose Antonio, Vinicio Horacio Santiago Cortes, Rita S. Rindermann, and Victor Herminio Palacio Muñoz. 2001. *El mercado del trigo en México antes TLACAN*. Centro de Investigaciones Economicas, Sociales y Teconologicas de la Agroindusria y la Agrícultura Mundial. Mexico: CIESTA, M-PIAI.

Bacon, David. 2013. *The Right to Stay Home: How US Policy Drives Mexican Migration*. Boston: Beacon Press.

Baldacci, Emanuele, Luiz de Mello, and Gabriela Inchauste. 2002. "Financial Crises, Poverty, and Income Distribution." *Finance and Development: A Quarterly Magazine of the IMF* 39(2): 1–8.

Bancomer. 2004. *Informe Económico* (October). Mexico City: Bancomer Grupo Financiero.

Barajas Duran, Rafael. 2008. *Cómo la hacen de Pemex: La nueva guerra del petróleo*. Mexico City: Editorial Planeta.

Barnes, Guillermo. 1992. *Lessons from Bank Privatization in Mexico*. Financial Policy & Systems Division, Country Economics Department. WPS no. 1027. Washington, DC: World Bank.

Bean, Frank, and Lindsay Lowell. 2007. "Unauthorized Migration." In *The New Americans: A Guide to Immigration since 1965*, ed. Mary C. Waters and Reed Ueda, 70–82. Cambridge, MA: Harvard University Press.

Beaumier, Guy. 1990. *Free Trade in North America: The Maquiladora Factor*. Ottawa, ON: Research Branch, Canada Library of Parliament, Depository Services Program, BP-247E Report.

Blauner, Robert. 1994 [1972]. "Colonized and Immigrant Minorities." In *From Different Shores: Perspectives on Race and Ethnicity in America*, ed. Ronald Takaki, 149–160. New York: Oxford University Press.

Boué, Juan Carlos. 2006. Aspectos fiscales de la apertura petrolera en México. In *Hacia la integracion de los mercados petroleros en America?* ed. Isabelle Rousseau, 341–375. Mexico City: El Colegio de México.

Boughton, James. 2001. *The Silent Revolution: The International Monetary Fund 1979–89*. Washington, DC: IMF.

Brown, Jonathan. 1993. *Oil and Revolution in Mexico*. Berkeley: University of California Press.

———, and Alan Knight, eds. 1992. *The Mexican Petroleum Industry in the Twentieth Century*. Austin: University of Texas Press.

Burdeau, Cain. 2015. "Appeals Court Hears Arguments on Obama Immigration Action." ABC News, April 17. http://abcnews.go.com/US/wireStory/appeals-court-obamas-immigration-action-30384504.

Burnstein, John. 2007. *US-Mexico Agricultural Trade and Rural Poverty in Mexico*. Report from a Task Force Convened by the Woodrow Wilson Center's Mexico Institute and Fundacion IDEA. Washington, DC: Woodrow Wilson International Center for Scholars.

Calderón, Esteban B. 1975. *Juicio sobre la guerra del Yaqui y génesis de la Huelga de Cananea*. Mexico City: Centro de Estudios Historicos del Movimiento Obrero Mexicano (CEHSMO).

Callahan, James M. 1932. *American Foreign Policy in Mexican Relations*. New York: MacMillan Co.

Cañas, Jesus, and Roberto Coronado. 2002. "Maquiladora Industry: Past, Present and Future." *Business Frontier*, no. 2. Federal Reserve Bank of Dallas. http://www .dallasfed.org/assets/documents/research/busfront/bus0202.pdf.

———, and Robert W. Gilmer. 2007. "Economic Trends in the Desert Southwest: Mexico Regulatory Change Redefines Maquiladora." *Crossroads*, no. 1. Federal Reserve Bank of Dallas. http://www.dallasfed.org/assets/documents/research/cross roads/2007/cross0701c.pdf.

Cárdenas Garcia, Jaime. 2009. *En defensa del petróleo*. Mexico City: Instituto de Investigaciones Jurídicas, UNAM.

Cardoso, Fernando H. 1977. "El consumo de la teoria de la dependencia en los Estados Unidos." *El Trimestre Economico*, 173 (44/1) Enero Marzo: 33–52.

———, and Enzo Faletto. 1979. *Dependency and Development in Latin America*. Trans. Marjory Mattingly Urquidi. Berkeley: University of California Press.

Carton de Grammont, Hubert. 2000. "Política Neoliberal, estructura productiva y organización social de los productores una visión de conjunto." In *Los pequeños productores rurales en México: Las reformas y las opciones*, ed. Antonio Yuñez-Naude, 73–99. Mexico City: Centro de Estudios Económicos, El Colegio de México.

Caruso-Cabrera, Michelle. 2014. "Oil Is Falling, but This Big Producer Isn't Worried. "*CNBC*, December 11. http://www.cnbc.com/id/102260792.

Castañeda, Jorge G. 2007. *Ex Mex: From Migrants to Immigrants*. New York: New Press.

Castillo, E. Eduardo. 2013. "Mexican Senate Passes Oil Reform and Discusses Privatization." *Huffington Post*, December 11. http://www.huffingtonpost.com/2013/12 /12/mexican-senate-oil-reform_n_4432345.html.

Cattan, Nacha, and Eric Martin. 2013. "Mexico Lower House Approves Tax Law Adding Junk Food Levy." *Bloomberg News*. http://www.bloomberg.com/news /articles/2013-10-17/mexico-congress-panel-advances-tax-bill-as-junk-food -duty-added.

Chávez Ramírez, Paulina. 1996. *Las cartas de intención y las politicas de estabilización y ajuste estructural de México: 1982–1994*. Mexico City: Instituto de Investigaciones Económicas, UNAM.

CNN. 2012. "López Obrador Demands Recount in Mexican Election Vote." *CNN: International*, July 3. http://www.cnn.com/2012/07/03/world/americas/mexico -elections.

Cockcroft, James D. 1986. *Outlaws in the Promised Land: Mexican Immigrant Workers and America's Future*. New York: Grove Press, Inc.

Coerver, Don, and Linda Hall. 1984. *Texas and the Mexican Revolution: A Study in State and National Border Policy 1910–1920.* San Antonio: Trinity University Press.

Corbacho, Ana, and Gerd Schwartz. 2002. "Mexico: Experiences with Pro-Poor Expenditure Policies." International Monetary Fund Working Paper, WP/02/12. Washington, DC: IMF.

Cordera, Rolando, and Carlos Tello. 2010. *México: La disputa por la nación perspectivas y opciones de desarollo.* Mexico City: Grupo Editorial Siglo Veintiuno.

Craig, Richard. 1971. *The Bracero Program: Interest Groups and Foreign Policy.* Austin: University of Texas Press.

Davids, Jules. 1976. *American Political and Economic Penetration of Mexico, 1877–1920.* New York: Arno Press.

Díaz Serrano, Jorge. 1992. *La privatización del petróleo mexicano.* Mexico City: Editorial Planeta.

———. 1989. *Yo, Jorge Díaz Serrano.* Mexico City: Editorial Planeta Mexicana.

Dinero. 2013. "INEGI advierte que seis de cada 10 empleos son informales en México." *NOTimex/Dinero*, June 18. http://www.dineroenimagen.com/2013-06-18/21844.

Domínguez, Jorge, and Rafael Fernández de Castro. 2009. *The United States and Mexico: Between Partnership and Conflict.* 2d ed. New York: Routledge.

Doyle, Michael. 2007. "Farmers Might Add Guest-Worker Plan to Farm Bill." McClatchyDC. http://www.mcclatchydc.com/2007/10/31/20985/lawmakersmight-add-guest-worker.html#.UipFEBxiDXE.

Durand, Jorge. 2007. "The Bracero Program (1942–1964): A Critical Appraisal." *Migración y Desarrollo*, Second Semester, 25–40.

El Economista. 2013. "Aumenta clase media en México: INEGI." June 12. http://www.eleconomista.com.mx/finanzas-publicas/2013/06/12/aumenta-clase-media-mexico-inegi.

Edelman, Marc, and Angelique Haugerud. 2004. *The Anthropology of Development and Globalization.* Malden, MA: Blackwell.

Escobar, Arturo. 2012. New Preface in *Encountering Development: The Making and Unmaking of the Third World.* Princeton, NJ: Princeton University Press.

———. 2010. "Latin America at a Crossroads." *Cultural Studies* 24(1): 1–65.

———. 1995. *Encountering Development: The Making and Unmaking of the Third World.* Princeton, NJ: Princeton University Press.

Escobar Latapí, Agustín, and Mercedes González de la Rocha. 1995. "Crisis, Restructuring and Urban Poverty in Mexico." *Environment and Urbanization* 7(1): 57–78.

Esteva, Gustavo. 1987. "Food Needs and Capacities: Four Centuries of Conflict." In *Food Policy in Mexico: The Search for Self-Sufficiency*, ed. James Austin and Gustavo Esteva, 23–47. Ithaca, NY: Cornell University Press.

Excelsior. 2013. "Reforma energética: Cárdenas ve aún possible revertir reforma energética." December 12. http://www.excelsior.com.mx/nacional/2013/12/13/933686.

Fabens, Isabella. 2013. *China's Latest Investments in Mexico: The Plight of Maquiladora Workers.* Council on Hemispheric Affairs. http://www.coha.org/chinas-latest-investments-in-mexico-the-plight-of-maquiladora-workers/.

Fabian, Johannes. 2002. *Time and the Other: How Anthropology Makes Its Object.* 2d ed. New York: Columbia University Press.

Falk, Pamela. 1987. *Petroleum and Mexico's Future.* Boulder, CO: Westview Press.

Flores-Macias, Francisco José. 2010. "Explaining the Behavior of State-Owned Enterprises: Mexico's Pemex in Comparative Perspective." PhD diss., Massachusetts Institute of Technology.

Flores-Quiroga, Aldo. 1998. *Proteccionismo versus librecambio: La economía política de la protección comerical en México 1970–1994.* Mexico City: Fondo de Cultura Economica.

Forbes. 1999. "A Decade of Wealth." October 11. http://www.Forbes.com/forbes /1999/0705/6401154a.html.

Forbes Mexico. 2014. A cuál clase social perteneces? May 17. http://www.forbes.com .mx/sites/a-cual-clase-social-perteneces-segun-la-se.

Fox, Jonathan. 1992. *The Politics of Food in Mexico: State Power and Social Mobilization.* Ithaca, NY: Cornell University Press.

Frank, Andre Gunder. 1967. *Capitalism and Underdevelopment in Latin America: Historical Studies of Chile and Brazil.* New York: Monthly Review Press.

Fredrickson, George. 2002. *Racism: A Short History.* Princeton, NJ: Princeton University Press.

Fundación Colossio. 2008. *La reforma energética balance y retos.* Mexico City: Fundacion Colossio.

Galarza, Ernesto. 1977. *Tragedy at Chualar. El crucero de las treinta dos cruces.* Santa Barbara, CA: McNally and Loftin.

———. 1964. *Merchants of Labor: The Mexican Bracero Story.* Santa Barbara, CA: McNally and Loftin.

Gallegos, Raul. 2013. "How Mexico Can Restrike Oil." *Bloomberg News,* December 17. http://www.bloombergview.com/articles/2013-12-17/how-mexico-can -restrike-oil.

Garfield, Chester, Joseph Gilbert, and Ezekiel Pogue. 1921. *America's Power Resources.* New York: Century.

Gilly, Adolfo. 1994. *La revolucion interrumpida.* Edición Corregida y Aumentada. Mexico City: Edicion, Era, Coleción Problemas de Mexico.

Girón González, Alicia. 2001. "Deuda externa." In *Financias publicas de México,* ed. Enrique Arriaga Concha, 241–285. Mexico City: Instituto Politecnico Nacional.

Gomez, Alan, and David Jackson. 2015. "Deportation Program Still Stalled by Courts." *USA Today,* May 27, p. 5A.

González, Nayeli. 2013. "Ejerce Pemex 40% de su presupuestos para 2013." March 9. http://www.milenio.com/politica/Ejerce-Pemex presupesto_0_146985656.html.

Gonzalez, Richard. 1985. "Reducing Energy Uncertainties Essential for Adequate Supplies." In *Improving U.S. Energy Security,* ed. Richard Gonzalez, Raymond Smilor, and Joel Darmstadter, 53–70. Cambridge, MA: Ballinger Pub. Co.

———, Raymond Smilor, and Joel Darmstadter. 1985. *Improving U.S. Energy Security.* Cambridge, MA: Ballinger.

González Ramirez, Manuel. 1986. *La revolución social de México, Vol. 1. Las ideas—la violencia.* Mexico City: Fonda de Cultura Económica.

———. 1974. *La Huelga de Cananea.* Mexico City: Fondo de Cultura Económica.

———. 1941: *El petróleo Mexicano: La expropiación petrolera ante el derecho internacional.* Mexico City: Editorial America.

Goodsell, James Nelson. 1981. "Mexico's Oil Minister Resigns over Oil-Price." *Christian Science Monitor,* June 8. http://www.csmonitor.com/1981/0608/060875.html.

Grayson, George. 1980. *The Politics of Mexican Oil.* Pittsburgh: University of Pittsburgh Press.

Grebler, Leo, Joan Moore, and Ralph Guzman. 1970. *The Mexican American People: The Nation's Second Largest Minority.* New York: Free Press.

Guerrero Andrade, Manuel. 2005. *De la gestión estatal al mercado global: Los sistemas de intervención estatal en la comercialización del maíz en México, 1936 to 2000.* Mexico City: UNAM-Xochimilco.

Guthrie, Amy. 2010. "Mexico Supreme Court Allows New Private Service Oil Contracts." *Wall Street Journal,* December 8. http://online.wsj.com/article/SB1000142 405274870349350457600778186313572.html.

Hanchard, Jenna. 2013. "Secy. of Agriculture Tom Vilsack Talks Immigration Reform at the American Royal in Kansas City." *KSHB 41,* June 21. http://www.kshb.com /news/secy-of-agriculture-tom-vilsack-talks-immigration-reform-at-the-american -royal-in-kansas-city.html.

Harl, Neil. 1990. *The Farm Debt Crisis of the 1980s.* Ames: Iowa State University Press.

Harris, Marvin. 1968. *The Rise of Anthropological Theory: A History of Theories of Culture.* New York: Crowell.

Harvey, David. 2006. "Neo-Liberalism and the Restoration of Class Power." In *Spaces of Global Capitalism: Towards a Theory of Geographical Development,* 9–68. New York: Verso.

Harvey, Neil. 1998. *The Chiapas Rebellion: The Struggle for Land and Democracy.* Durham, NC: Duke University Press.

Herskovits, Melville. 1938. *Acculturation: The Study of Cultural Contact.* New York: J. J. Augustin.

Hobson, John. 1902. *Imperialism: A Study.* London: James Nisbet and Co.

Hoffman, Abraham. 1974. *Unwanted Mexican Americans in the Great Depression: Repatriation Pressures, 1929–1939.* Tucson: University of Arizona Press.

IEA. 2015. "FAQ: Oil." International Energy Agency. http://www.iea.org/aboutus/faqs /oil.

Iglesias Prieto, Norma. 2001. *Beautiful Flowers of the Maquiladora: Life Histories of Women Workers in Tijuana.* Austin: University of Texas Press.

Iliff, Laurence. 2013. "Pemex's Proven Oil Reserves Edge Up to 13.87 Billion Barrels." *Wall Street Journal,* March 17. http://online.wsj.com/article/BT-CO-20130317-70 1705.html.

IMF. International Monetary Fund, IMF Archive documents of Mexico, Western Hemispheric Department. Washington, DC: IMF.

———. Executive Board Documents. EBD/82/99 (1982), Mexico economic adjustment program for 1982, April 26; EBD/94/200 (1994), Mexico-Press Communiqué, public finances, December 21, 1994.

―――. Executive Board Specials. EBS/90/10 (1990), Mexico extended arrangement, January 17, 1990.

―――. Board Document—Staff Memoranda. SM/94/41 (1994), Statistical Annex, February 10, 1994; SM/94/31 (1993), Consultation Article 4, corrected; SM/94/270 (1994), Adoption of indirect instruments of monetary policy, November 17, 1994.

―――. Executive Board Minutes. EBM/93/11 (1994), Consultation Article 4, February 23, 1994, corrected version of EBM93/11 (1993).

―――. BUFF/ED: Statements by Executive Directors at Executive Board Meetings. BUFF/ED/94/27 (1994), Statement of Mr. Marino, February 25 and 28, 1994, recovery overview.

―――. Country Reports. CR/13/334 (2013), Mexico: Article IV Consultation; CR/12/317 (2012), Article IV Consultation; CR/10/70 (2010) (labor market informality); CR/11/250 (2011), Mexico: Article IV Consultation; CR/9/7 (2009) (crime and banking); CR/6/352 (2006), Mexico: Article IV Consultation; CR/5/427 (2005), Mexico: Article IV Consultation; CR/4/418 (2004), Mexico Selected Issues (economic collapses); CR/4/419 (2004), Mexico: Article IV Consultation.

―――. 1982. *Annual Report 1982: International Monetary Fund.* Washington, DC: IMF.

Inda, Jonathan Xavier. 2007. "The Value of Immigrant Life." In *Women and Migration in the U.S.-Mexico Borderlands: A Reader,* ed. Denise A. Segura and Patricia Zavella, 134–157. Durham, NC: Duke University Press.

International Institute of Sustainable Development. 2007. "Mexican Government Approves Farm Subsidies as NAFTA Barriers Are Set to Expire." Global Subsidies Initiative, International Institute of Sustainable Development. March 29. http://www.iisd.org/gsi/news/mexican-government-approves-farm-subsidies-nafta-barriers-set-expire.

Jayne, Catherine. 2001. *Oil, War, and Anglo-American Relations: American and British Reactions to Mexico's Expropriation of Foreign Oil Properties, 1937–1941.* Westport, CT: Greenwood Press.

Jimenez, Eugenia. 2012. "Descarta Madero apoyar agenda de López Obrador." *Milenio,* August 14, p. 4.

Kahn, Carrie. 2013. "How NAFTA Helped the Mexican Billionaires' Club." *National Public Radio,* December 31. http://www.npr.org/2014/01/01/258668414/how-nafta-helped-the-mexican-billionaires-club.

Kluger, Richard. 2007. *Seizing Destiny: How America Grew from Sea to Shining Sea.* New York: Alfred A. Knopf, Inc.

Knight, Alan. 1986. *The Mexican Revolution.* Vol. 1: *Porfirian Liberals and Peasants.* New York: Cambridge University Press.

Kose, Ayhan, Guy Meredith, and Christopher Towe. 2004. *How Has NAFTA Affected the Mexican Economy? Review and Evidence.* International Monetary Fund Working Paper, WP/04/59. Washington, DC: IMF.

Krauss, Clifford. 1997. "Mexican Data Suggest 30% Overstatement of Oil Reserves." *New York Times*, March 18. http://www.nytimes.com/1997/03/18/business/mexican-data-suggest-30-overstatement-of-oil-reserves.html.

Kroll, Luisa. 2012. "Forbes World's Billionaire List." *Forbes*. http://www.forbes.com/sites/luisakroll/2012/03/07/forbes-worlds-billionaires-2012.

Kubbah, Abdul Amir. 1974. *OPEC Past and Present*. Report no. 18463, OPEC Information Department. Vienna, Austria: OPEC.

Larsen, Donald. 1993. "Policies for Copy with Price Uncertainties for Mexican Maize." Policy Research Writing Papers, WPS 1120, Inter Trade, International Economics Department. Washington, DC: International Monetary Fund.

Leos, Raul Muñoz. 2006. *Pemex en la encrucijada: Rencuento de una gestion*. Mexico City: Nuevo Siglo Aguilar.

Lesser, Alexander. 1933. *The Pawnee Ghost Dance Hand Game: A Study of Cultural Change*. Vol. 26, Columbia University Contributions to Anthropology. New York: Columbia University Press.

Lewis, Arthur. 1954. "Economic Development with Unlimited Supplies of Labour." *Manchester School of Economics and Social Studies* 22: 139–191.

López, Ann Aurelia. 2007. *The Farmworkers' Journey*. Berkeley: University of California Press.

López Obrador, Andrés. 2010. *La mafia que se adueño de México y el 2012*. Mexico City: Grijalbo.

Lorey, David. 1999. *The U.S.-Mexican Border in the Twentieth Century: A History of Economic and Social Transformation*. Wilmington, DE: Scholarly Resources.

Lustig, Nora. 1995. "Mexico: The Social Costs of Adjustment." In *Privatization amidst Poverty: Contemporary Challenges in Latin American Political Economy*, ed. Jorge Lawton, 69–109. Boulder, CO: Lynne Rienner.

Manzo Yépez, José Luis. 1996. *Que hacer con PEMEX? Una alternative a la privatización*. Mexico City: Grijalbo.

Martin, Eric, and Carlos Manuel Rodriguez. 2012. "Pemex Reform Harder as PRI Misses Majority in Congress." *Bloomberg*, July 3. http://www.bloomberg.com/news/2012-07-03/pemex-overhaul-harder-as-pri-misses-mexican-congress-majority.html.

Martin, Philip, and David Martin. 1994. *The Endless Quest: Helping America's Farm Workers*. Boulder, CO: Westview Press.

Martínez, José. 2011. *Los secretos del hombre mas rico del mundo Carlos Slim*. Mexico City: Oceano Exprés.

Martínez Fernández, Raymundo, comp. 1996. *Deuda externa vs. desarollo economico: Analysis y síntesis de la Reunion Internacional sobre Deuda Externa y Alternativas*. Mexico City: Foro de Apoyo Mutuo.

Mattar, Jorge. 1998. Export Promotion in Mexico. *Integration and Trade*, no. 4/5, January-August, 179–217.

Menchaca, Martha. 2011. *Naturalizing Mexican Immigrants: A Texas History*. Austin: University of Texas Press.

———. 2001. *Recovering History, Constructing Race: The Indian, Black, and White Roots of Mexican Americans*. Austin: University of Texas Press.

————. 1995. *Mexican Outsiders: A Community History of Marginalization and Discrimination in California*. Austin: University of Texas Press.

Mendez, Enrique, and José Antonio Román. 2014. "Mexico Chamber of Deputies Approves Hydrocarbon Laws." Mexico Voices: Address Mexico's Challenges. July 29. http://mexicovoices.blogspot.com/2014/07/mexicos-chamber-of-deputies-approves.html.

Mercado, Angelica. 2012. "No habremos de aceptar argucias legaloides: AMLO." *Milenio*, August 13, p. 4.

Meyer, C. Michael, William L. Sherman, and Susan M. Deeds. 2007. *The Course of Mexican History*. 8th ed. New York: Oxford University Press.

Meyer, Lorenzo. 1972. *México y Los Estados Unidos en el Conflicto Petrolero (1917–1942)*. Mexico City: El Colegio de Mexico.

Miller, Robert R. 1985. *Mexico: A History*. Norman: University of Oklahoma Press.

Miller, Tom. 1981. *On the Border: Portraits of America's Southwestern Frontier*. New York: Harper and Row.

Mines, Richard, and Ricardo Anzaldúa. 1982. *New Migrants vs. Old Migrants: Alternative Labor Market Structures in the California Citrus Industry*. Monograph 9, Program in United States Mexican Studies. San Diego: University of California, San Diego.

Mintz, Sidney. 1985a. *History, Evolution, and the Concept of Culture: Selected Papers by Alexander Lesser*. New York: Cambridge University Press.

————. 1985b. *Sweetness and Power: The Place of Sugar in Modern History*. New York: Viking.

Morales, Isidro. 1992. "The Consolidation and Expansion of Pemex, 1947 to 1958." In *The Mexican Petroleum Industry in the Twentieth Century*, ed. Jonathan Brown and Alan Knight, 208–232. Austin: University of Texas Press.

Moreno-Brid, Juan Carlos, and Jaime Ros. 2009. *Development and Growth in the Mexican Economy: A Historical Perspective*. New York: Oxford University Press.

MSNBC. 2014. "John Boehner: President Obama Is 'Acting Like' a King, Emperor." Video, November 20. http://www.msnbc.com/msnbc/watch/boehner-obama-is-acting-like-an-emperor-361510979671.

Murillo, Oscar Cuevas. 2008. La reforma liberales, materia de propiedad según Wistano Luis Orozco y Andrés Molina Enriquéz. *Investigaciones Jurdicas de UNAM*, 97–128. Zacatecas, Mexico: Universidad Autonoma de Zacatecas. http://www.jurdicas.unam.mex.

Nasar, Faisal Adel. 1981. "Opec's and Mexico's Petroleum in the Context of Economic Development: Competition or Cooperation." PhD diss., University of Texas at Austin.

National Immigration Law Center. 2015. "National Immigration Law Center Applauds Justice Department Decision to Seek Supreme Court Review in *Texas v. U.S.*" http://www.nilc.org/nr111015.html.

Negroponte, Diana Villiers. 2014. "Mexico's Secondary Law Provides a Path Forward for New Investments in the Hydrocarbon Sector." *Brookings Institution*, June 25. http://www.brookings.edu/research/articles/2014/06/25-mexico-law-hydrocarbons-sector-negroponte.

Nevins, Joseph. 2002. *Operation Gatekeeper: The Rise of the "Illegal Alien" and the Making of the U.S.-Mexico Boundary.* New York: Routledge.

Ngai, Mae M. 2004. *Impossible Subjects: Illegal Aliens and the Making of Modern America.* Princeton, NJ: Princeton University Press.

Nielsen, Lynge. 2011. *Classifications of Countries Based on Their Level of Development: How It Is Done and How It Could Be Done.* International Monetary Fund Working Paper, WP/11/31. www.imf.org/external/pubs/ft/wp/2011/wp1131.pdf.

Nieto, Francisco, and Horacio Jiménez. 2014. Alistan consulta popular en torno a ley energética. *El Universal,* April 25. http://www.eluniversal.com.mx/nacion -mexico/2014/impreso/alistan-consulta-popular-en-torno-a-ley-energetica -215061.html.

O'Brien, Michael. 2012. "Obama Administration Won't Seek Deportation of Young Illegal Immigrants." *NBC News,* June 15. http://nbcpolitics.nbcnews.com/_news /2012/06/15/12238794-obama-administration-wont-seek-deportation-of-young-il legal-immigrants?lite.

Oleaga, Michael. 2015a. "Immigration Reform News Update: Obama Meets with Latino, Immigrants Rights Advocates for Private Meeting." *Latin Post,* February 26. http://www.latinpost.com/articles/39942/20150226/immigration-reform-news -update-obama-meets-latino-immigrants-rights-advocates.htm.

———. 2015b. "Immigration Reform Lauwsuit: Obama, Justice Department Will Not Issue Appeal to Supreme Court." *Latin Post,* May 29. http://www.latinpost.com /articles/56428/20150529/immigration-reform-lawsuit-obama-justice-department -will-issue-appeal-supreme.htm.

O'Neil, Shannon. 2013. *Two Nations Indivisible: Mexico, the United States, and the Road Ahead.* New York: Oxford University Press.

Ong, Aihwa. 2007. "Introduction: Neoliberalism as Exception, Exception to Neoliberalism." In *Neoliberalism as Exception: Mutations in Citizenship and Sovereignty,* 1–27. Durham, NC: Duke University Press.

OPEC. 2015. "OPEC Basket Prices, April, 23, 2015" (online interactive). Organization of the Petroleum Exporting Countries. Vienna, Austria. http://www.opec.org /opec-web/en/index.htm.

———. 2014. "OPEC Basket Prices, March 12, 2014" (online interactive). Organization of the Petroleum Exporting Countries. Vienna, Austria. http://www.opec .org/opec_web/en/data_graphs/40.htm.

———. 2013. *OPEC Annual Statistical Bulletin 2013.* Organization of the Petroleum Exporting Countries. http://www.opec.org/opec_web/static_files_ project/media /downloads/publications/ASB2001pdf.

———. 2001. *OPEC Annual Statistical Bulletin 2001.* Organization of the Petroleum Exporting Countries. http://www.opec.org/opec_web/static_files_project/media /downloads/publications/ASB2001pdf.

———. 1999. *OPEC Annual Statistical Bulletin 1999.* Organization of the Petroleum Exporting Countries. http://www.opec.org/opec_web/static_files_project/media /downloads/publications/ASB1999pdf.

Orozco, Wistano Luis. 1895. *Legeslación y jurisprudencia sobre terrenos baldíos*. Mexico: Impresa de el Tiempo.

Ortiz Santillán, José Luis. 2013. "México, mas pobre al iniciar el Siglo XXI." *Effeta.info*, June 16. http://www.effeta.info/blog/2013/06/mexico-mas-pobre-al -iniciar-el-siglo-xxi.

Padgett, Tim. 2012. "How Enrique Peña Nieto Won Himself and His Party the Mexican Presidency." *Time*, July 2. World edition. http://world.time.com/2012/07/02 /mexico-election-how-enrique-pena-nieto-won-himself-and-his-party-the -presidency.

Padilla, Liliana. 2012. "Presenta AMLO su decálogo parea invalidar elección." *Milenio*, August 8, p. 4.

Perez, Ana Lilia. 2010. *Camisas azules, manos negras: El saqueo de Pemex desde Los Pinos*. Mexico City: Random House, Mondadori, S.A. de C.V.

Perez-Rocha, Manuel, and Emily Schwartz Grego. 2008. "Mexico's Oil Referendum." *Foreign Policy in Focus*, August 6. Institute for Policy Studies. http://fpif.org/mexi cos_oil_referendum/.

Pletcher, David. 1958. "The Fall of Silver in Mexico, 1870–1910, and Its Effect on American Investments." *Journal of Economic History* 18(1): 33–55.

Prebisch, Raúl. 1971. *Change and Development—Latin America's Great Task: Report Submitted to the Inter-American Development Bank*. New York: Praeger Publishers.

———. 1950. *The Economic Development of Latin America and Its Principal Problems*. United Nations Economic Commission for Latin America. Lake Success, NY: United Nations Department of Economic Affairs.

Puyana Mutis, Alicia. 2006. "Las Fuerzas que molden la política energetica Mexicana: Entre la constitución y el TLCAN." In *¿Hacia la Integracion de los Mercados Petroleros en America?* ed. Isabelle Rousseau, 257–298. Mexico City: El Colegio de Mexico.

Ramirez, Miguel. 1989. *The IMF Austerity Program, 1983–87: Miguel de la Madrid's Legacy*. Occasional Papers in Latin America, no. 7. Center for Latin American Studies, Brown University.

Randall, Laura. 1989. *The Political Economy of Mexican Oil*. New York: Praeger.

Relinger, Rick. 2010. "NAFTA and U.S. Corn Subsidies: Explaining the Displacement of Mexico's Corn Farmers." *Prospect: Journal of International Affairs at UCSD*, April 19. http://prospectjournal.org/2010/04/19/nafta-and-u-s-corn-sub sidies-explaining-the-displacement-of-mexicos-corn-farmers-2.

Restuccia, Daniella. 2013. "New Era for Mexico: Congress Passes Energy Reform Bill." *Rio Grande Digital.com*, December 18. http://www.riograndedigital.com/2013 /12/18/new-era-for-mexico-congress-passes-energy-reform-bill/.

Reuters. 2014. "Update 1—Mexico's Senate Committees Approve Energy Reform Bills." *Reuters Edition*. http://www.reuters.com/article/2014/07/15/mexico-reforms -idUSL2N0PQ21U20140715.

Revueltas, Roman. 2012. "¿Como se va a 'legitimar' Peña-Nieto?" *Milenio*, August 10, p. 3.

Rippy, Merrill. 1972. *Oil and the Mexican Revolution*. Leiden, Netherlands: E. J. Brill.

Rodriguez, Carlos Manuel, and Jonathan Roeder. 2012. "A Big Oil Find May Derail Reforms in Mexico." *Bloomberg Businessweek*, October 4. http://www.bloomberg .com/bw/articles/2012–10–04/a-big-oil-find-may-derail-reforms-in-mexico.

Romero, Matias. 1889. "The Annexation of Mexico." *North American Review* 148 (390): 525–537.

Ros, Jaime, and César Bouillon. 2001. "La liberálización de la balanza de pagos en México: Effectos en el crecimiento, la desigualdad y la pobreza." In *Liberálización, desigualdad, y pobreza: America Latina y el Caribe en los 90s*, ed. Enrique Ganuza, 714–765. United Nations: CEPAL.

Rothenberg, Daniel. 2000. *With These Hands: The Hidden World of Migrant Farmworkers Today*. Berkeley: University of California Press.

Samora, Julian. 1970. *Los Mojados: The Wetback Story*. Notre Dame, IN: University of Notre Dame Press.

Santiago, Myrna. 2006. *The Ecology of Oil: Environment, Labor and the Mexican Revolution, 1900–1938*. New York: Cambridge University Press.

Schatan, Jacobo. 1987. "SAM's Influence in Food Consumption and Nutrition." In *Food Policy in Mexico: The Search for Self-Sufficiency*, ed. James Austin and Gustavo Esteva, 249–259. Ithaca, NY: Cornell University Press.

Scholes, Walter. 1957. *Mexican Politics during the Juarez Regime 1855–1872*. Columbia: University of Missouri Press.

Shabad, Rebecca, and Cristina Marcos. 2014. "House Passes Bill to Defund Obama's Immigration Order." *The Hill*, January 14. http://thehill.com/blogs/floor-action /house/229469-house-votes-to-defund-obamas-immigration-orders.

Shao, Stephen Pinyee. 1956. "Coal and Its Trend in the United States." PhD diss., University of Texas at Austin.

Sheingate, Adam. 2001. *The Rise of the Agricultural Welfare State: Institutions and Interest Group Power in the United States, France, and Japan*. Princeton, NJ: Princeton University Press.

Shields, David. 2005. *PEMEX La Reforma Petrolera*. Mexico City: Editorial Planeta.
————. 2003. *PEMEX: Un Futuro Incierto*. Mexico City: Editorial Planeta.

Shipway, Martin. 2008. *Decolonization and Its Impact: A Comparative Approach to the End of the Colonial Empires*. Malden, MA: Wiley-Blackwell.

Shulman, Robert, comp. 1986. *Laws Affecting Farm Employment in California*. North Highland, CA: Cooperative Extension, University of California Division of Agricultural Resources.

Sloan, John W. 1978. "United States Policy Responses to the Mexican Revolution: A Partial Application of the Bureaucratic Politics Model." *Journal of Latin American Studies* 10(2): 283–308.

Society of Petroleum Engineers. 2005. *Comparison of Selected Reserves and Resource Classifications and Associated Definitions*. Oil and Gas Reserves Committee, "Mapping" Subcommittee, Final Report—December 2005. http://www.spe.org/indus try/docs/OGR_Mapping.pdf#redirected_from=/industry/reserves/mapping.php.

Sosnick, Stephen. 1978. *Hired Hands: Seasonal Farm Workers in the United States*. Santa Barbara, CA: McNally and Loftin, West.

Spicer, Edward. 1981. *Cycles of Conquest: The Impact of Spain, Mexico, and the United States on the Indians of the Southwest, 1533–1960*. Tucson: University of Arizona Press.

Stephen, Lynn. 2007. *Transborder Lives: Indigenous Oaxacans in Mexico, California, and Oregon*. Durham, NC: Duke University Press.

Stevenson, Mark. 2013. "Mexico State Legislatures Pass Energy Reform Bill." *Huffington Post*, December 17. http://www.huffingtonpost.com/2013/12/17/mexico-energy-reform-bill_n_4458549.html.

Stukey, Elizabeth Smith. 1984. "Economic Interdependence and the Sovereignty of Debtor Nations: A Comparison of Mexican and Argentine Reactions to International Monetary Fund Stabilization." *Fordham International Law Journal* 8(3): 543–588.

Takaki, Ronald. 2000 (1979). *Iron Cages: Race and Culture in Nineteenth-Century America*. New York: Alfred A. Knopf.

Tanfani, Joseph. 2015. "Obama Adjusts His Legal Strategy in Immigration." *Tribune News Service*, May 29. http://www.governing.com/topics/politics/tns-obama-immigration-legal-strategy.html.

Taush, Arno. 1993. *Towards a Socio-Liberal Theory of World Development*. London: Macmillan Press.

Taylor, Ronald. 1975. *Chavez and the Farm Workers: A Study in the Acquisition and Use of Power*. Boston: Beacon Press.

TeleSUR. 2014. "Mexican High Court Dismisses Energy Reform Referendum." *TeleSUR*, October 24. http://www.telesurtv.net/english/news/Mexican-High-Court-Dismisses-Energy-Reform-Referendum-20141030-0046.html.

Tello, Carlos. 2009. *Sobre la desigualdad en México*. Mexico City: Facultad de Economia, UNAM.

Tiano, Susan. 2006. "The Changing Gender Composition of the Maquiladora Workforce along the U.S.-Mexico Border." In *Women and Change at the U.S.-Mexico Border: Mobility, Labor, and Activism*, ed. Doreen J. Mattingly and Ellen R. Hansen, 73–90. Tucson: University of Arizona Press.

Twin Plant News. 2008. "What Is a Maquila?" http://www.twinplantnews.org.

Univision. 2013. "We Are Not Equal." Pt. 1. http://feeds.univision.com/openpage/2013-07 18/especial-inmigracion-mobile.

Vásquez, Josephina Zorida, and Lorenzo Meyer. 1985. *The United States and Mexico*. Chicago: University of Chicago Press.

Vera, Manuel, and Andrew Farris. 2014. "Mexican President Signs Historic Energy Reform Law." *Bracewell and Giulani* (blog), August 18. http://www.energylegalblog.com/archives/2014/08/18/5779.

Wallerstein, Immanuel. 1974. *The Modern World-System*. Vol. 1, *Capitalist Agriculture and the Origins of the European World-Economy in the Sixteenth Century*. New York: Academic Press.

Weber, David. 1982. *The Mexican Frontier, 1821–1846: The American Southwest under Mexico*. Albuquerque: University of New Mexico Press.

Wells, Miriam. 1981. "Oldtimers and Newcomers: The Role of Context in Mexican American Assimilation." *Aztlan* 7: 267–290.

White, Leslie. 1949. *The Science of Culture, a Study of Man and Civilization.* New York: Grove Press.

Williams, Adam, Eric Martin, and Nacha Cattan. 2013. "Mexico Passes Oil Bill Seen Luring $20 Billion a Year." *Bloomberg*, December 12. http://www.bloomberg.com /news/2013–12–12/mexico-lower-house-passes-oil-overhaul-to-break-state -monopoly.html.

Wilson, Patricia. 2010. "Maquiladoras." *Texas State Historical Association.* http://www .tshaonline.org/handbook/online/articles/dzm02.

Wolf, Eric. 1982. *Europe and the People without History.* Berkeley: University of California Press.

Woolley, John T., and Gerhard Peters. 2008. "927—Statement on Signing the Immigration and Nationality Act Amendments of 1976." Gerald Ford, October 21. The American Presidency Project (website). Santa Barbara, CA: University of California. http://www.presidency.ucsb.edu/ws/index.php?pid=6495&st=&st1=.

World Bank. 2013. "Mexico: Data." http://data.worldbank.org/country/mexico.

———. 2012. World Data Bank: World Development Indicators. http://databank. worldbank.org/data//reports.aspx?source=2&country=MEX&series=&period=.

———. 2005. Mexico: *Income Generation and Social Protection for the Poor: Executive Summary.* No. 36853. http://documents.worldbank.org/curated/en/2005/01/6954 135/mexico-income-generation-social-protection-poor.

———. 1980. *Report and Recommendation of the President of the International Bank for Reconstruction and Development to the Executive Directors on a Proposed Loan to Nacional Financiera. S.A. with the Guarantee of United Mexican States for a Small and Medium-Scale Mining Development Project,* Report no. P-2735, March 6. http://www-wds.worldbank.org/external/default/WDSContentServer/WDSP /IB/2000/06/02/000178830_98101902344345/Rendered/INDEX/multi_page.txt.

Zabludovsky, Jaime Enrique. 1998. "La deuda exterior publica." In *Un siglo de deuda pública en México,* ed. Leonor Ludlow and Carlos Marichal, 152–189. Instituto de Investigaciones Históricas-UNAM. Mexico City: UNAM.

Zalloum, Abdulhay Yahya. 2007. *Oil Crusades: America through Arab Eyes.* Ann Arbor, MI: Pluto Press.

Zavella, Patricia. 2011. *I'm Neither Here nor There: Mexicans' Quotidian Struggles with Migration and Poverty.* Durham, NC: Duke University Press.

Zhou, Moming. 2015. "Oil Jumps $60 for First Time This Year as Glut Eases." *BloombergBusiness,* May 5. www.bloomberg.com/news/articles/2015–05–05/u-s-oil -rises-to-60-for-first-time-this-year-as-glut-recedes.

Zuñiga, Juan Antonio. 2013. "Están en clase baja 66.4 millones de mexicanos, reporta el Inegi." *La Jornada,* June 13. http://www.jornada.unam.mx/2013/06/13 /economia/025n1eco.

GOVERNMENT DOCUMENTS

Baker, Bryan, and Nancy Rytina. 2013. *Estimates of the Unauthorized Immigrant Population Residing in the United States: January 2012.* US Department of Homeland Security, Office of Immigration Statistics, Policy Directorate, March 2013. https://www.dhs.gov/sites/default/files/publications/ois_ill_pe_2012_2.pdf.

Commission on Agricultural Workers. 1993. *The Winter Vegetable Industry in South Texas.* Executive Summary, prepared for the Commission on Agricultural Workers 1989–1993. Texas State Government. Austin, TX: State Printers.

Congressional Record. 2007. Senate, 110th Cong., 1st sess., vol. 153, pt. 2.

———. 1986a. 99th Cong., 2d sess., Daily Record no. 140, October 10.

———. 1986b. 99th Cong., 2d sess., vol. 132, no. 97.

———. 1982a. 97th Cong., 2d sess., vol. 128, pt. 18.

———. 1982b. 97th Cong., 2d sess., vol. 128, pt. 17.

———. 1982c. 97th Cong., 2d sess., vol. 128, pt. 16.

———. 1973a. 93rd Cong., 1st sess., vol. 119, pt. 30.

———. 1973b. 93rd Cong., 1st sess., vol. 119, pt. 29.

Consejo Nacional de Población. 2009. *Informe ejecución del Programa de Acción de la Conferencia Internacional sobre la Población y el Desarollo 1994–2009.* Mexico City: CONAPO.

Cruz Miramontes, Rodolfo, Oscar Cruz Barney, and Patricia Aguilar Mendez. 2009. *Elementos juridicos: Para una controversia comerical contra del maiz.* Mexico: LX Legislatura/Congreso de la Union.

Derecho Internacional Mexicano. 1877. Vol. 1, Tratados y Convenciones Celebrados y Ratificados por la Republica Mexicana. Jose Fernandez, comp. Mexico: Imprenta de Gonzalo A. Esteva.

Executive Agreement Series. EAS 1858. Migratory Workers. *Treaties and Other International Agreements of the United States of America 1776–1949,* 9:1224–1232. Entered into force April 9, 1947.

———. EAS 1684. Migratory Workers. *Treaties and Other International Agreements of the United States of America 1776–1949,* 9:1215–1220. Entered into force November 16, 1946.

———. EAS 278. Migratory Workers. *Treaties and Other International Agreements of the United States of America 1776 to 1949,* 9:1069–1075. Entered into force August 4, 1942.

EIA (US Energy Information Administration). 2015a. "International Energy and Data Analysis: Total Petroleum Consumption." EIA, Independent Statistics and Analysis. http://www.eia.gov/beta/international/index.cfm?topl=con.

———. 2015b. "How Much Petroleum Does the U.S. Import and from Where?" EIA, Independent Statistics and Analysis. http://www.eia.gov/tools/faqs/faq.cfm?id=727&t=6.

———. 2014a. "Petroleum and Other Liquids." EIA, Independent Statistics and Analysis. http://www.eia.gov/dnav/pet/hist/LeafHandler.ashx?n=pet&s=rwtc&f=a.

———. 2014b. "The Availability and Price of Petroleum and Petroleum Products in Countries Other Than Iran." February 20 update. EIA, Independent Statistics and Analysis. http://www.eia.gov/analysis/requests/ndaa/.

―――. 2014c. "Mexico: Overview Data, Proved Oil Reserves, Exports." EIA, Independent Statistics and Analysis. http://www.eia.gov/countries/country-data .cfm?fips=MX.

―――. 2014d. "United States: Overview Data, Proved Oil Reserves." EIA, Independent Statistics and Analysis. http://www.eia.gov/countries/country-data .cfm?fips=US.

―――. 2013a. "Today in Energy, Concentration of U.S. Crude Oil Imports among Top Five Suppliers Highest since 1997." EIA, Independent Statistics and Analysis. http://www.eia.gov/todayinenergy/detail.cfm?id=10911.

―――. 2013b. "International Petroleum Consumption." EIA, Independent Statistics and Analysis. http://www.eia.gov/cfapps/ipdbproject/iedindex3.cfm?tid=5&p id=5&aid=2&cid=regions&syid=2006&eyid=2010&unit=TBPD.

―――. 2013c. "U.S. Crude Oil Proved Reserves (Million Barrels)." EIA, Independent Statistics and Analysis. http://www.eia.gov/dnav/pet/hist/LeafHandler.ashx ?n=PET&s=RCRR01NU S_1&f=A.

―――. 2013d. "Mexico International Statistics: Crude Oil Proved Reserves." EIA, Independent Statistics and Analysis. http://www.eia.gov/dnav/pet/hist/Leaf Handler.ashx?n=PET&s=RCRR01NUS_1&f=A.

―――. 2013e. "The Availability and Price of Petroleum and Petroleum Products in Countries other than Iran." November 2-December 13 update, p. 5. EIA, Independent Statistics and Analysis. http://www.eia.gov/analysis/requests/ndaa.

―――. 2012. "The Availability and Price of Petroleum and Petroleum Products in Countries other than Iran." February 29 update, p. 3. EIA, Independent Statistics and Analysis. http://www.eia.gov/analysis/requests/ndaa.

―――. 2011. "Mexico: Analysis Brief." EIA, Independent Statistical Analysis, Official Energy Statistics from the US Government. http://www.eia.gov/countries /cab.cfm?fips=MX.

Employment Development Department Research Division. 1986. *Employment Development Department Oxnard-Ventura Labor Market Bulletin*. Los Angeles: Employment Data and Research Division.

Estados Unidos Mexicanos, Presidencia de la República. 2014a. Enrique Peña Nieto, Presidente de los Estados Unidos Mexicanos, a sus habitantes sabed: Que la Comisión Permanente del Honorable Congreso de la Unión, se ha servido dirigirme el siguiente: Decreto "La Comisión Permanente del Honorable Congreso de la Unión, en uso de la facultad que le confiere el Articulo 135 Constitucional y previa la aprobación de las Cámaras de Diputados y de Senadores del Congreso General de los Estados Unidos Mexicanos, Así como la mayoría de las Legislaturas de los Estados, declara reformadas y adicionadas diversas disposiciones de la Constitución Política de los Estados Unidos Mexicanos, en materia energía." (Mexico's Energy Reform Decree, Article 25, 27, 28). http://www.dof.gob.mx/nota_detalle .php?codigo=5327463&fecha=20/12/2013.

―――. 2014b. Iniciativas de Leyes Secundarias. Presidente de la mesa directive: De la Cámara de Senadores del H. Congreso de la Unión Presente Presidente, Enrique Peña Nieto, April 28. http://cdn.reformaenergetica.gob.mx/1-ley-de-hidrocarburos .pdf.

GAO (US Government Accountability Office). 2008. *Oil and Gas Royalties: The Federal System for Collecting Oil and Gas Revenues Needs Comprehensive Reassessment.* GAO-08-691.

———. 2007. *Crude Oil: Uncertainty about Future Oil Supply Makes It Important to Develop a Strategy for Addressing a Peak and Decline in Oil Production.* GAO-07283.

———. 1997. *Financial Crisis Management: Four Financial Crises in the 1980s.* GGD-9796.

———. 1996. *Mexico's Financial Crisis: Origins, Awareness, Assistance, and Initial Efforts to Recover.* GGD-96-56.

Gobierno de la República de México. 2014a. *Reforma Energética: Mexico.* http://refor mas.gob.mx/wp-content/uploads/2014/04/Explicacion_ampliada_de_la_Reforma _Energetica1.pdf.

———. 2014b. *Reforma energética: México.* http://www.presidencia.gob.mx/reforma energetica/#!landing.

Hoefer, Michael, Nancy Rytina, and Bryan Baker. 2011. *Estimates of the Unauthorized Immigrant Population Residing in the United States: January 2010.* US Department of Homeland Security, Office of Immigration Statistics, Policy Directorate, Population Estimates, February 2011.

Immigration from Countries of the Western Hemisphere: Hearings before the Committee on Immigration and Naturalization, House of Representatives. 1927. 70th Cong., 1st sess., no. 70.1.5.

INEGI (Instituto Nacional de Estadísticas y Geografía). 2014. *Estadisticas.* (Current national statistics of the population, online.) http://www.inegi.org.mx/est/conte nidos/proyectos/estadistica/default.aspx.

———. 2013a. *Anuario estadístico de los Estados Unidos Mexicanos, 2012.* Aguascalientes, Mexico: INEGI.

———. 2013b. *Clases medias en México.* Boletín de Investigacion 256/13. Aguascalientes, Mexico: INEGI.

———. 2012a. *Anuario estadístico de los Estados Unidos Mexicanos, 2011.* Aguascalientes, Mexico: INEGI.

———. 2012b. *México de un Vistazo.* Aguascalientes, Mexico: INEGI.

———. 2011. *Sistema de cuentas nacionales de México.* Aguascalientes, Mexico: INEGI.

———. 2009a. *Mexico hoy* [Mexico Today]. Aguascalientes, Mexico: INEGI.

———. 2009b. *Encuestra nacional de empleo en 2009.* Aguascalientes, Mexico: INEGI.

———. 2009c. *Resumen de los resultados de los censos económicos 2009.* Aguascalientes, Mexico: INEGI.

———. 2000a. *El sector alimentario en México.* Aguascalientes, Mexico: INEGI.

———. 2000b. *México en el Siglo XX.* Aguascalientes, Mexico: INEGI.

———. 1999. *Estadisticas del comercio exterior de México* 9, no. 12. Aguascalientes, Mexico: INEGI.

———. 1996. *Mexican Bulletin of Statistical Information*, no. 18. Aguascalientes, Mexico: INEGI.

———. 1984. *Encuestos Nacionales de Ingresos y Gastos de los hogares, cuarto tremestre.* Aguascalientes, Mexico: INEGI.

Kandel, William. 2008. *Profile of Hired Farmworkers: A 2008 Update.* US Department of Agriculture, Economic Research Service, ERR no. 60, July. http://www.UDSA.gov.

KAV 4268. February 21, 1995. Mexico: Guarantee Agreement. Signed February 21, entered into force February 21. *U.S. Treaties and other International Agreements.* http://www.heinonline.org.

KAV 4269. February 21, 1995. Mexico Economic Stabilization: Framework Agreement for Mexican Economic Stabilization, with Annexes. Signed February 21, entered into force February 21. *U.S. Treaties and other International Agreements.* http://www.heinonline.org.

KAV 2553. March 23, 1990. Swap Agreement among the U.S. Treasury and the Banco de México, Government of Mexico, Memorandum of Understanding. Signed March 23, entered into force March 23. *U.S. Treaties and other International Agreements.* http://www.heinonline.org.

Levine, Linda. 2009. *Farm Labor Shortage and Immigration Policy.* Congressional Research Service Reports. CRS RL 30395. Washington, DC: National Council for Science and the Environment.

———. 2004a. *Immigration: The Labor Market Effects of a Guest Worker Program for U.S. Farmers.* Congressional Research Service Reports. CRS Report 97–712E. Washington, DC: US Commission on Agricultural Workers.

———. 2004b. *Farm Labor Shortages and Immigration Policy.* Congressional Research Service Reports. CRS Report 30395. Washington, DC: US Commission on Agricultural Workers.

Marshall, Ray. 2013. "United States Department of Labor Office of the Assistant Secretary for Administration and Management. The Labor Department in the Carter Administration: A Summary Report—January 14, 1981." US Department of Labor, Employment Standards Administration. http://www.dol.gov/oasam/programs/history/carter-esa.htm.

México Presidencia de la Republica. 2014. Salarios Mínimos 2013. Gobierno Mexicano. http://www.presidencia.gob.mx/salarios-minimos-2013.

NAWS (National Agricultural Workers Survey). 2005. *Findings from the National Agricultural Workers Survey (NAWS 2001–2002).* A Demographic and Employment Profile of United States Farmworkers, March 2005, Research Report no. 9. US Department of Labor, Office of Program Economics. http://www.doleta.gov/agworker/report9/naws_rpt9.pdf.

———. 2000. *Findings from the National Agricultural Workers Survey (NAWS 1997–1998).* A Demographic and Employment Profile of United States Farmworkers, Research Report no. 8. US Department of Labor, Office of Program Economics. http://www.doleta.gov/agworker/naws.cfm.

Pemex. 2015a. Precios, April 23. http://www.gas.pemex.com/PGPB/Productos+y +servicios/Gas+licuado/Precios/.

———. 2015b. *Monthly Petroleum Statistics: Average Realized Price of Crude Oil Exports.* January and February. http://www.ri.pemex.com/files/dcpe/petro/eprecio promedio_ing.pdf.

———. 2014a. *Monthly Petroleum Statistics: Average Realized Price of Crude Oil Exports.* March 12. http://www.ri.pemex.com/files/dcpe/petro/epreciopromedio _ing.pdf.

———. 2014b. *Composicion de las 145 regiones de precios,* July 18. http://www.gas .pemex.com/PortalPublico/EstadosMunicipiosGL.aspx?.

———. 2014c. Palabras del Director General de Pemex, Emilio Lozoya, Durante la Presentactión de la Ronda Cero y Uno de la Reforma Energetica. August 13. http://www.pemex.com/saladeprensa/discursos/Paginas/discurso_dg_ronda cero_140813.aspx.

———. 2014d. *Petróleos Mexicanos,* Primer informe trimestral 2014, Articulo 71 (párrafo primero), Ley de Petróleos Mexicanos. May. http://www.pemex.com /acerca/informes_publicaciones/Documents/Articulo%2071/Primer_Informe _Trimestral_2014.pdf.

———. *Statistical Yearbook* (Anuario Estadístico) (1977–2013). Mexico City: Pemex. http://www.pemex.com/en/investors/publications/Paginas/statistical-yearbook .aspx.

———. *Informe Anual* (1997, 1998, 2003, 2008, 2012). http://www.pemex.com/en/ investors/publications/Paginas/annual-report.aspx.

———. *Memoria de labores* (1978, 1979, 1989, 1990). http://www.pemex.com/acerca /informes_publicaciones/Paginas/memoria_labores.aspx#.UqeV-xxiAno.

———. 1988. *La industria petrolera en México: Cronología 1857–1988.* Gloria Villegas Moreno, Coordinación General.

———. 1958. *Los Veinte Años de la Industria Petrolera Nacional: Informes del 18 de Marzo 1938–1958.* No. 35578.

PMI. 2013. "¿Que PMI? Petróleos Mexicanos Comercio Internacionales." *Pemex,* GobMX. http://portal.pmi.com.mx/Paginas/QuienesSomos.aspx?IdSec=13.

Rytina, Nancy, and Selena Caldera. 2007. *Naturalizations in the United States: 2007.* Department of Homeland Security, Office of Immigration Statistics, Policy Directorate. http://www.dhs.gov.immigrationstatistics.

Secretaria de Desarrollo Agrario, Territorial y Urbano. 2014. El Presidente Enrique Peña Nieto Promulga las Leyes Secundarias de la Reforma Energética. Nota no. 046, Govierno Mexicano, SEDATU. http://www.sedatu.gob.mx/sraweb/pie_nota /pie-de-foto-2014/pie-agosto-2014/19723/.

SCHP (Secretaría de Hacienda y Crédito Público). 2014. Poder Executivo: Ley de Ingresos Sobre Hidrocarburos. Diario Oficial. 1st ed. August 11. http://www.dof .gob.mx/nota_to_doc.php?codnota=5355982.

Seidband, Debbie. 2004. "U.S. Wheat and Corn Exports to Mexico Thrive under NAFTA." *AgExporter* 16, no. 1: 1. http://permanent.access.gpo.gov/lps19216/www .fas.usda.gov/info/agexporter/2004/January.

Simanski, John. 2008. SAW data. E-mail to author, December 12. Department of Homeland Security, Office of Immigration Statistics, Policy Directorate.

Stana, Richard. 1999. *Unauthorized Employment of Aliens*. US General Accounting Office, Administration of Justice Division. Government Report 99–33. Washington, DC: GPO.

Statistical Yearbook of the Immigration and Naturalization Service 1994. Immigration and Naturalization Service, US Department of Justice. Washington, DC: GPO.

TIAS 10961. "Mexico Finance: Consolidation and Rescheduling of Certain Debts." *U.S. Treaties and Other International Agreements* 35, no. 4. *UST* 4609, May 2, 1984.

TIAS 5492. "Migrant Workers — Mexican Agricultural Workers." *U.S. Treaties and Other International Agreements* 14, no. 2. *UST* 1804, December 20, 1963.

TIAS 5311. "Migrant Workers — Mexican Agricultural Workers." *U.S. Treaties and Other International Agreements* 14, no. 1. *UST* 307, January 10 and February 25, 1963.

TIAS 5160. "Mexican Agricultural Workers." *U.S. Treaties and Other International Agreements* 13, no. 2. *UST* 2022, December 29, 1961.

TIAS 4815. "Mexican Agricultural Workers." *U.S. Treaties and Other International Agreements* 12, no. 1. *UST* 1081, June 27, 1961.

TIAS 4374. "Mexican Agricultural Workers." *U.S. Treaties and Other International Agreements* 10, no. 2. *UST* 2036, October 23, 1959.

TIAS 3242. "Agricultural Workers: Recommendation by Joint Migratory Labor Commission." *U.S. Treaties and Other International Agreements* 6, no. 1. *UST* 1017, April 14, 1955.

TIAS 3054. "Mexican Agricultural Workers." *U.S. Treaties and Other International Agreements* 5, no. 2. *UST* 1793, August 6, 1954.

TIAS 3043. "Mexican Agricultural Workers." *U.S. Treaties and Other International Agreements* 5, no. 2. *UST* 1669, July 16, 1954.

TIAS 2928. "Mexican Agricultural Workers." *U.S. Treaties and Other International Agreements* 5, no. 1. *UST* 353, December 30, 31, 1953.

TIAS 2260. "Mexican Agricultural Workers." *U.S. Treaties and Other International Agreements* 2, no. 1. *UST* 1048, August 1, 1949.

US Census. 1962. *Current Population Reports Consumer Income*. Series P-60, no. 37. Washington, DC: GPO.

———. 1854. *Statistical View of the United States Census; Being a Compendium of the Seventh Census*. Washington, DC: GPO.

US Census of Agriculture. 2015. *The Faces of U.S. Agriculture*. USDA. National Agricultural Statistics Service. http://www.agcensus.usda.gov/Partners/Infographics /Faces_of_Agriculture.pdf.

US Congressional Serial Set. 1954a. House of Representatives, Doc. no. 1199, 83d Cong., 2d sess.

———. 1954b. House of Representatives, Doc. no. 1436–9 1954, Hearing ID: HRG-1954-HAG-0001, 83d Cong., 2d sess.

———. 1931a. House of Representatives, Doc. no. 715, 71st Cong., 3d sess.

———. 1931b. House of Representatives, Doc. no. 2710, 71st Cong., 3d sess.

————. 1929. House of Representatives, Doc. no. 2418, 70th Cong., 2d sess.

————. 1926. Senate Doc. no. 96, 69th Cong., 1st sess.

USDA (US Department of Agriculture). 2015a. "Farm Income and Wealth Statistics." USDA, Economic Research Service. http://www.ers.usda.gov/data-products/farm-income-and-wealth-statistics/government-payments-by-program.aspx.

————. 2015b. "Immigration and the Rural Workforce." USDA, Economic Research Service. http://www.ers.usda.gov/topics/in-the-news/immigration-and-the-rural-workforce.aspx.

————. 2015c. "The Evolution of US Agricultural Exports over the Last Two Decades." USDA, Economic Research Service. http://www.ers.usda.gov/topics/international-markets-trade/us-agricultural-trade/exports/interactive-chart-the-evolution-of-us-agricultural-exports-over-the-last-two-decades.aspx.

————. 2008. "Fact Sheet: North American Free Trade Agreement (NAFTA), January 2008." *USDA*, Foreign Agricultural Service. http://www.fas.usda.gov/sites/development/files/nafta1.14.2008_0.pdf.

————. 2004. "U.S.-Mexico Corn Trade during the NAFTA Era: New Twists to an Old Story." *USDA*, Electronic Outlook Report from the Economic Research Service. http://www.ers.usda.gov/media/168930I/fds04d0I.pdf.

US Department of Energy. 2014. "Strategic Petroleum Reserve." http://www.energy.gov/fe/services/petroleum-reserves/strategic-petroleum-reserve.

US Department of Homeland Security. 2013. Written Testimony of US Department of Homeland Security Janet Napolitano for a Senate Committee on the Judiciary hearing titled "The Border Security, Economic Opportunity, and Immigration Modernization Act, S.744." April 23. http://www.dhs.gov/news/2013/04/23/written-testimony-dhs-secretary-janet-napolitano-senate-committee-judiciary-hearing.

————. 2012. "Executing Prosecutorial Discretion with Respect to Individuals Who Came to the United States as Children, June 15, 2012." Memorandum, Janet Napolitano, Secretary of Homeland Security. http://www.dhs.gov/xlibrary/assets/s1-exercising-prosecutorial-discretion-individuals-who-came-to-us-as-children.pdf.

US Department of Justice. 1971. "Immigrants Admitted, by Country of Origin of Birth, Years Ended June 30, 1962–1971" (Table 14). Office of Immigration Statistics, Immigration and Naturalization Service, Department of Homeland Security. http://www.dhs.gov.

US Department of Labor. 2009. "History of Federal Minimum Wage Rates under the Fair Labor Standards Act, 1938–2009." Wage and Hour Division. http://www.dol.gov/whd/minwage/chart.htm.

————. 2002. Table 1: Employment status of the civilian noninstitutional population, 1940 to date. http://www.bls.gov/cps/aa2002/cpsaat1.pdf.

————. 1988. "Wage and Hour Division: History of Changes to the Minimum Wage Law. In *Minimum Wage and Maximum Hours Standards under the Fair Labor Standards Act, 1988 Report to the Congress under Section 4(d)(1) of the FLSA*. Wage and Hour Division. http://www.dol.gov/whd/minwage/coverage.htm.

————. 1964a. *Hired Farmworkers in the United States*. Bureau of Employment Secu-

rity. Washington, DC: GPO. Wirtz Labor Library, Digital Archives, http://www
.dol.gov.

———. 1964b. *Fifty-Second Annual Report U.S. Department of Labor.* Washington,
DC: GPO. Wirtz Labor Library, Digital Archives, http://www.dol.gov.

———. 1963. *United States Department of Labor Annual Report, 1963.* Washington,
DC: GPO. Wirtz Labor Library, Digital Archives, http://www.dol.gov.

———. 1957a. *Proceedings of Consultation on Migratory Labor.* Church Councils,
Labor Statistics. Wirtz Labor Library, Digital Archives, http://www.dol.gov/oasam
/ibrary/digital.

———. 1957b. *Summary of the Labor Studies in Mexico.* Bureau of Labor Statistics.
Wirtz Labor Library, Digital Archives, http://www.dol.gov/oasam/ibrary/digital.

US Department of State. 2014. "U.S. Relations with Mexico." Bureau of Western
Hemisphere Affairs, September 10. http://www.state.gov/r/pa/ei/bgn/35749.htm.

———. 2013. "Mexico." In *Treaties in Force: A List of Treaties and Other International
Agreements of the United States in Force on January 1, 2013,* 192–193. http://www
.state.gov/documents/organization/218912.pdf.

US House Resolution 1773, Agricultural Guestworker Act (AG Act). 2014. Library of
Congress, 113th Cong., 2013–2014. https://www.congress.gov/bill/113th-congress
/house-bill/1773/all-info.

US Senate Act 744. 2013. *The Border Security, Economic Opportunity, and Immigration
Modernization Act, 2013.* Library of Congress, 113th Cong., 2013–2014. https://
www.congress.gov/bill/113th-congress/senate-bill/744.

US Senate Foreign Relations Committee. 2006. *Energy Security and Oil Dependence:
Hearing for the Committee on Foreign Relations U.S. Senate.* 109th Cong., May 16.
Washington, DC: Committee on Foreign Relations. http://www.gpoaccess.gov
/cong/index.html.

US Senate Judiciary Committee. 2013a. *Bipartisan Framework for Comprehensive Immigration Reform.* http://www.flake.senate.gov/documents/immigration_reform
.pdf.

———. 2013b. *Senate Judiciary Committee, Hearing on Immigration, April 23, 2013,
9 to 12 am.* C-Span. http://www.c-span.org/video/?312302–1/dhs-sec-napolitano
-testifies-immigration-reform.

Ventura County Agricultural Association. 1986. "Farm Labor Wage Survey in Ventura County." Survey conducted by Wally Haven, director of the Ventura County
Agricultural Association. Available at the Ventura County Agricultural Association, California.

White House. 2013. State of the Union, January 29. "Blueprint for Reform," in *Creating an Immigration System for the 21st Century.* http://www.whitehouse.gov
/issues/immigration/streamlining-immigration.

———. 2006. "President Bush Meets with President Fox in Cancun, Mexico."
March 30. Office of the US Press Secretary. http://georgewbush-whitehouse
.archives.gov/news/releases/2006/03/20060330–7.html.

US STATUTES

110 US Statutes at Large [pt. 4] (1996)
108 US Statutes at Large [pt. 5] (1994)
100 US Statutes at Large [pt. 4] (1986)
90 US Statutes at Large [pt. 2] (1976)
79 US Statutes at Large (1965)
76 US Statutes at Large (1962)
66 US Statutes at Large (1952)
65 US Statutes at Large (1951)

LEGAL CASES

Emanuel Braude et al., Appellant v. W. Willard Wirtz, Secretary of Labor, US, et al., Appellee, 350 Federal Reporter, 2d 702 (1965).
Silva v. Bell, 65 Federal Reporter, 2d 978 (1979).
State of Texas v. United States of America, US District Court for the Southern District of Texas Brownsville Division, Civil no. B-14-254 (2015). https://www.documentcloud.org/documents/1668197-hanen-opinion.html.
State of Texas, et al. v. United States, Appellant's Emergency Motion for Stay Pending Appeal, no. 15–40238 (2015). http://www.justice.gov/sites/default/files/opa/press-releases/attachments/2015/03/12/stay_motion_filed_0.pdf.

PRINT NEWSPAPERS

Dallas Morning Star, 1921
La Prensa, 1921
Milenio, 2012
USA Today, 2015

Index

Abbott, Greg, 141
AgJob Bills, 104–105, 132–134, 199n14
agrarian land reform policies: Article 27, 43, 45; of Cárdenas administration (agrarian reform), 62; and foreign ownership, 56–57; and privatization under President Salinas, 124–125
Agricultural Workers Act (HR 1773), 139, 141
Alemán, Miguel (Mexican president), 74–75, 77
alien land laws (Mexico), 54–57, 60. *See also* Cárdenas, Lázaro; Díaz, Porfirio; Land Law Act of 1883; Land Law of 1890; Mexican Revolution; Mineral Law of 1892; Mining Law of 1909
American Farm Bureau Federation, 101, 102, 104
Appadurai, Arjun, 17
Article 23 (Mexican Constitution), 197n3
Article 25 (Mexican Constitution), 183, 201n19
Article 27 (Mexican Constitution): and Calderón administration energy reforms, xii–xiii, 145, 149, 159, 169–170, 201n20; and Calles administration, 55–57; and ejidos, 124; enactment of (1917), 43–45; and energy land laws of 2014, 185; Mexican Supreme Court ruling on constitutionality

of, 46–47, 53, 55–56, 60; and Obregón administration, 50–54; and Peña-Nieto administration energy reforms, xiii, 145, 149, 176, 178–184, 201nn18–19, 201n22; and Salinas administration energy reforms, 117, 154–156; and Zedillo administration energy reforms, 164–165
Article 28 (Mexican Constitution), 145, 180–181
asymmetrical economic codependency: and Alemán administration, 75; and aspects of capitalism, 189–191; and Camacho administration, 67; and Cárdenas administration, 59; and de la Madrid administration, 111–113; and López Portillo administration, 106–110; and Obregón administration, 54; and Porfirian period, 20, 22; and Salinas administration, 115–120; and Second World War politics, 70–77; theory, xi–xii, xiv–xvi, 1–8. *See also* oil conditionality agreements; US conditionality agreements

Baker Plan, 111, 113, 199n4
Barajas Duran, Rafael, 159–160
Border Protection, Antiterrorism, and Illegal Immigration Control Act of 2005 (HR 4437), 132–133